YOUNG SCIENTIST USA.
NATURAL SCIENCE

Y.S.

3702 W Valley HWY N
STE 204-31245
Auburn, WA
98001

http://www.YoungScientistUSA.com/

Printed in the United States of America

Lulu, 2014

ISBN 978-1-312-13256-6

Table of Contents

Physical Science

Earth Science

Life Science

PHYSICAL SCIENCE

Influence of the Rare-Earth Elements Concentration on Catalytic and Physicochemical Properties of Pentasils in Toluene Alkylation with Methanol

Talekh Gakhramanov

Ayub S. Mamedov

Eldar Akhmedov

Nargiz Akhmedova

Natavan Makhmudova

Baku State University, Baku, Azerbaijan

Abstract. *Physicochemical and catalytic properties of HZSM-5 modified with the rare-earth elements (REE) have been studied, in toluene alkylation with methanol. During the REE zeolite chemical modification, redistribution of acidic sites and reduction of the zeolite sorption capacity have been shown. As a result, Lewis and Broensted acidic sites ratio increases more intensively, and the zeolite channels narrow, causing the catalyst para-selectivity to increase.*

Keywords: *toluene alkylation, pentasils, para-selectivity, acidic sites, xylenes.*

Zeolites of ZSM-5 type are widely used as catalysts in various petrochemical processes [1, 2]. Their catalytic activity in conversion processes of hydrocarbons of different classes is determined by both the molecular-sieve and acidic properties [2, 3].

Regulating the acidic properties and porous structure by zeolites chemical modification with various modifiers allows para-selective catalysts to be obtained for aromatics alkylation and disproportionation processes [4–6].

The purpose of this work is to study the influence of the REE nature and concentration on physicochemical and catalytic properties of the zeolite of ZSM-5 type in toluene alkylation with methanol.

Experimental

High-silica zeolite of ZSM-5 type with molar ratio $SiO_2/Al_2O_3 = 33$ was investigated, which was converted into NH_4-form by ion exchange according to the technique described earlier [6]. Zeolite of N-form was obtained by the thermal decomposition of NH_4-form at 500 °C during 4 hours. The catalysts modified with 2.5–10.0% (m/m) REE were obtained by impregnating zeolites of N-forms with REE nitrate solution at 80 °C during 6 hours. The samples were air-dried during 16 hours, and then dried in an oven at 110 °C during 4 hours; and finally, they were ignited in a muffle furnace at 500 °C during 4 hours. Chemical and adsorption methods of analysis were used to investigate the catalysts. Acidic properties of the modified zeolites were studied by the thermal desorption of ammonia according to the technique described in work [7]. Experiments were performed in a flow unit with a fixed bed catalyst of 4 cm³ in the linear flow reactor under atmospheric pressure in the presence of hydrogen over the range of temperatures 300–400 °C, with feed space velocity of 1 h^{-1} and C_7H_8:CH_3OH:H_2 molar ratio of 2:1:2. The reaction products were analyzed using chromatography [5].

Results and discussion

Incorporating lanthanum and holmium by impregnating zeolites with the solutions of the corresponding metal nitrates, followed by the salt decomposition at 550 °C, substantially changes the properties of the catalysts: their activity declines and the

selectivity of p-xylene formation increases (Figure 1). With increasing of lanthanum and holmium concentration in HZSM-5 to 10.0% (m/m), selectivity to p-xylene increases to 57.5–61.0%, and the yield of C_9-C_{10} aromatic hydrocarbons falls down to 1.5–2.0% (m/m).

At lanthanum and holmium concentration of 10.0% (m/m), HZSM-5 activity declines almost in half. Increasing of lanthanum and holmium concentration to 5.0% (m/m) decreases the yield of xylenes by ~4.0% (m/m). However, in this case, selectivity to p-xylene is rather high: 58.5–61.5%.

Figure 2 shows selectivity to p-xylene vs. the nature of the modifier. Incorporating different modifiers (La, Yb, Ho, Lu) by impregnation into HZSM-5 content in the amount of 5.0% (m/m) substantially increases p-selectivity and significantly decreases the rate of side reactions. In case of modification, selectivity to p-xylene rises from 24.7% to 61.5% that is almost by a factor of 2.0–2.5.

Study of the influence of the modifiers nature and concentration on the pentasil acidic properties has shown that, in modification, strong proton-donating sites decline from E>130 kJ/mol (Table 1). Increasing of Yb, Ho and La concentration in the samples to 10.0% (m/m) decreases the concentration of strong proton-donating sites by a factor of ~6.

Table 1 demonstrates that, as a result of the REE pentasiles modification, the new stronger aprotonic acidic sites are formed rather than L-sites of

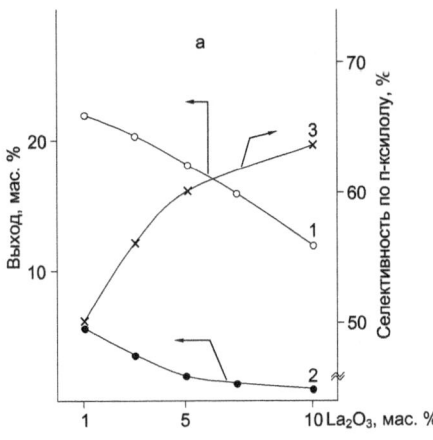 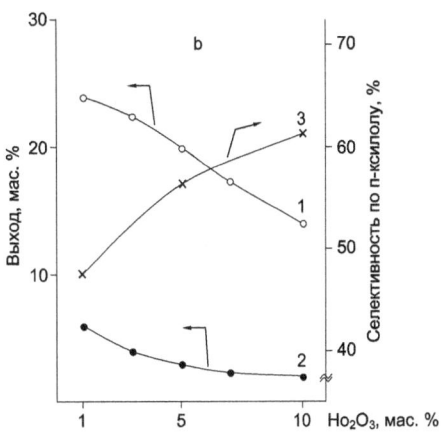

FIG. 1. The yield of xylenes (1), C_9-C_{10} aromatic hydrocarbons (2) and selectivity to p-xylene (3) vs. lanthanum (*a*) and holmium (*b*) concentration in HZSM-5 (350 °C, 2 h^{-1}, C_7H_8: CH_3OH=2:1)

Выход, мас. % — Yield, % (m/m)

Селективность по п-ксилолу, % — Selectivity to p-xylene, %

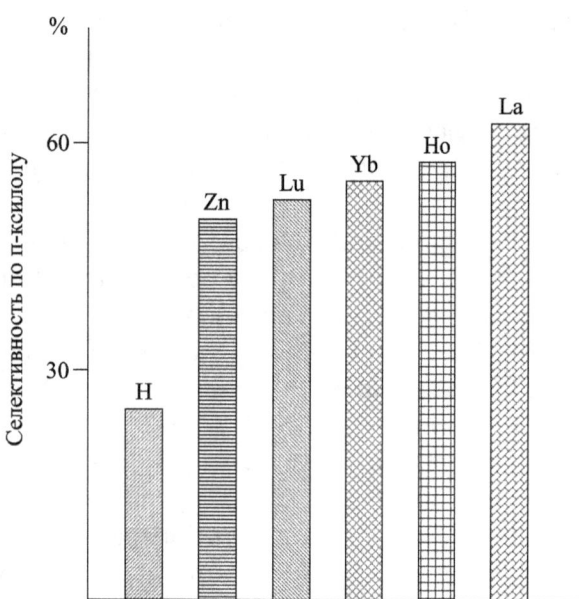

FIG. 2. Selectivity to p-xylene vs. the nature of the modifier (the content of the modifier in HZSM–5.0% (m/m))

Селективность по n-ксилолу — Selectivity to p-xylene

the original zeolites. Based on the results, it may be concluded that there is direct connection between the aprotonic acidity of the modified pentasiles and their activity and selectivity in toluene alkylation with methanol. As a result of modification, B to L sites ratio substantially changes: the incorporation of modifiers into pentasiles increases B/L sites ratio. With increasing of the modifiers content in zeolite to 5.0% (m/m), B/L sites ratio significantly rises.

With further increasing of the modifiers content, this ratio is little changed.

Thus, the catalytic data together with the results of acidity investigation leads to the conclusion that new L-sites are formed on the surface of decanted pentasiles of different composition as a result of modification. These sites are able to activate the toluene and methanol molecules, to prevent transformation of the resultant alkenes, disproportionation and transalkylation of aromatic hydrocarbons due to the electron-accepting properties of the aprotonic site, thereby increasing the catalyst para-selectivity.

The change in the acidic properties and selectivity of zeolites after their modification is explained as follows. In the process of N-pentasiles impregnation with REE^{+3} salt solution, the part of H^+ are replaced with REE^{+3} and/or $REE(OH)^{+2}$ that are formed in the hydrolysis of REE nitrate; and after the salt decomposition, REE_2O_3 is formed — the basic oxide that may interact with H^+-zeolite — the solid acid, according to the scheme

$$REE_2O_3 + 6\,H^+ \rightarrow 2\,REE^{+3} + 3\,H_2O$$

The part of REE_2O_3 remains in the channels and on the outer surface of aluminum silicate crystals, changing the sizes of the channels and their exit windows. As a result, the proton acidity and their activity in the carbonium-ion reactions declines; and the diffusion properties of contacts which are closely related to the catalysts para-selectivity are changed.

TABLE 1. Redistribution of the amount of acidic sites ($\mu mol \cdot g^{-1}$) over the activation energy values (E, kJ/mol) for the zeolites modified with different modifiers

Sample	E<95 kJ/mol	95≤ E<130 kJ/mol	E>130 kJ/mol	High temperature shoulder	130<E<(160–175) kJ/mol	I_{1448}/I_{1550} (L/B)
1% Yb-HZSM	380.2	226.5	167.5	59.3 (>160 °C)	108.2	1.4
3% Yb-HZSM	212.5	238.3	114.2	37.4 (>165 °C)	76.8	3.2
5% Yb-HZSM	220.8	258.5	82.2	31.9 (>170 °C)	50.3	3.8
10% Yb-HZSM	242.3	270.2	48.4	27.2 (>165 °C)	21.2	4.0
5% Ho-HZSM	213.4	265.2	77.3	26.1 (>170 °C)	51.2	4.0
10% Ho-HZSM	248.2	284.3	43.2	24.1 (>165 °C)	19.1	4.2

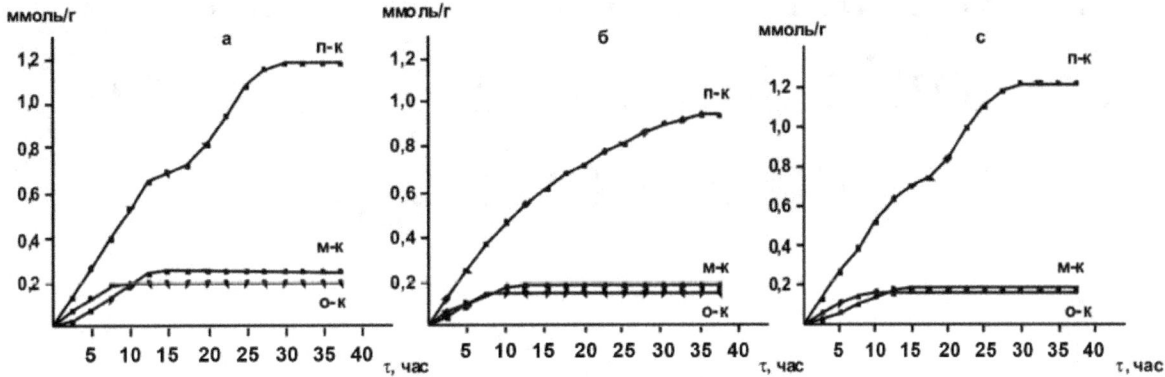

FIG. 3. The influence of holmium concentration on sorption capacity of HZSM-5 in relation to xylenes: a) 1% Ho–HZSM; б) 3.0% Ho–HZSM; с) 5.0% Ho–HZSM

ммоль/г — mmol/g; т, час — т, hours; М-К — m-xylene; О-К — o-xylene; П-К — p-xylene

The influence of modification and holmium content in HZSM-5 on its sorption capacity in relation to xylenes is shown in the Figure 3. During modification, the sorption capacity of the zeolite declines in relation to the m- and o-xylenes which are of the larger critical molecule diameter (0.7–0.75 nm) than p-xylene molecules (0.55 nm). Increasing of holmium concentration leads to further decrease in the sorption capacity of zeolite in relation to the m- and o-xylenes, which leads to an increase in p-xylene/m-xylene ratio.

Apparently the increase in the catalysts para-selectivity is associated with the weakening of the strong Broensted sites and narrowing of the zeolite channels as a result of modification. In this case, an increased selectivity to p-xylene may be explained by the higher diffusion rate in the channels of p-xylene pentasile, therefore p-xylene molecules are easier desorbed into gas phase than m- and o-xylene molecules are, which will be in the zeolite pores for a longer time and will undergo further isomerization.

References

1. Akpolat O., Gunduz G. Izomerization of m-xylene.// J.Appl.Scien. — 2005. —V 5 (2). — №2. — P. 236-248.
2. Y Chen Nai. Personal Perspective of the Development of para selective ZSM-5 catalysts.// Ind. Eng. Chem. Res. — 2001. —Vol. 40. — P. 4157-4161.
3. S.M. Csicsery. Spare selective catalysis.// Zeolites. — 1984. —V. 4. — P. 202-205.
4. A. Borgna, J. Sepúlveda, S.I. Magni, C.R. Apesteguía. Active sites in the alkylation of toluene with methanol: a study by selective acid-base poisoning. // Applied Catalysis: A-General. — 2004. —V.276. — P.207-215.
5. Vinek H., Derewinski M., Mirth G., Lercher J.A. Alkylation of toluene with methanol over alkali exchanged ZSM-5 // J. Appl. Catal. — 1991. —V.6. — P.277-284.
6. Aliev I.A., Akhmedov E.I., Mamedov E.S., Gakhramanov T.O. Toluene ethylation with ethanol on cadmium-promoted high silica zeolite.// Petroleum Chemistry. — 2010. —Vol. 50. —№. 5. — P. 373–375.
7. Yushchenko V.V. Calculation of the catalysts acidity spectra based on the temperature-programmed adsorption of ammonia.// Journal of Physical Chemistry. — 1997. —V. 71. — No 4. — P.28-32.

n-Heptane Isomerization over Bicationic Palladium-Containing Zeolites of Y-type

Ayten Mamedova

Sakina Mirzalieva

Eldar Akhmedov

Sabit Mamedov

Baku State University, Baku, Azerbaijan

Abstract: *Influence of polyvalent cations (Ho^{3+}, Cr^{3+}, Co^{2+}, Mn^{2+}) on catalytic and acidic properties of Pd/ CaY catalyst in n-heptane isomerization has been studied. High promoting effect of polyvalent cations on Pd-zeolite catalyst activity has been established; and polyvalent cations have been shown to influence the activity of Pd/CaY catalyst by regulating the amount and strength of acidic sites.*

Keywords: *zeolites, n-heptane isomerization, polyvalent cations, acidic sites.*

The nature of a polyvalent cation provides significant influence on the metal-zeolite catalysts activity in hydrocarbons transformation reactions occurring by carbonium-ion mechanism [1-3]. There are numerous literature works on investigating catalytic properties of the metal-zeolite catalysts with polyvalent cations in n-paraffins isomerization [3-6]. However, in n-paraffins isomerization, metal-zeolite catalysts containing two polyvalent cations have been little studied.

This paper provides the data on the influence of polyvalent cations (Ho^{3+}, Cr^{3+}, Co^{2+}, Mn^{2+}) on acidic and catalytic properties of Pd-zeolite catalyst (Pd/CaY) in n-heptane isomerization.

Experimental

Pd and MeCa forms of zeolite Y with SiO_2/Al_2O_3 molar ratio of 5 and 0.5% (m/m) palladium content have been examined as catalysts. The prepared catalysts contained 25% (m/m) Al_2O_3 as a binding agent. The technique for bicationic zeolites and catalysts preparation was the same as described in [2]. Ion exchange degree in Pd/MeCaY catalysts was Ca – 81.0%, Me – 5.0%.

n-Heptane isomerization was performed in a flow unit with catalyst loading of 5,0 cm^3 at the temperature of 553-603 K, H_2:C_7H_{16} molar ratio of 3, with n-heptane space velocity of 1 h^{-1}, under atmospheric pressure. The experiment lasted 30 minutes. The reaction products were analyzed by the chromatographic method [2].

The catalysts acidic properties were analyzed by the temperature-programmed desorption (TPD) of ammonia [8]. By this technique, the amounts ($\mu mol \cdot g^{-1}$) were determined, which correspond to the concentrations of acidic sites of different strength, according to our conventional classification of

TABLE 1. Influence of polyvalent cations incorporated into Pd/CaY catalyst content on the acidity spectrum

Catalysts	Amount of ammonia desorbed, $\mu mol \cdot g^{-1}$		
	573K	673K	773K
Pd/CaY	80	57	20
Pd/CrCaY	180	98	30
Pd/HoCaY	172	70	11
Pd/MnCaY	153	82	27
Pd/CoCaY	144	50	10

concentrations: middle (573K), strong (673K), and highly strong (773K). Acidity data for Pd-zeolite catalysts that were obtained by TPD are given in the table 1.

Figure 1 also demonstrates that incorporating Ho^{+3}, Mn^{+2}, Co^{+2} and Cr^{+3} into Pd/CaY catalyst content not only reinforces their activity, but also reduces the temperature of n-heptane isomerization reaction by 20-30 K.

From the table, it follows that the nature of a polyvalent cation incorporated into Pd/CaY catalyst content provides significant influence on its acidic properties. Replacing 5% Ca^{+2} cations with Ho^{+3} and Co^{+2} cations mainly increases the amount of middle acidic sites (from 80 to 160 $\mu mol \cdot g^{-1}$), while replacing the same amount of Ca^{+2} cation with Mn^{+2} and Cr^{+3} cations not only increases the amount of

middle, but also strong (from 60 to 92 $\mu mol \cdot g^{-1}$) and highly strong (from 20 to 33 $\mu mol \cdot g^{-1}$) acidic sites.

Higher selectivity of Pd-zeolite catalysts based on combination of Ca^{+2} cations with Ho^{+3} and Co^{+-2} cations appears to be related to the presence of small amount of highly strong acidic sites that are responsible for the occurrence of n-heptane cracking [2,4].

The results obtained suggest that catalytic n-heptane isomerization over polyvalent cation modified Pd-zeolite catalyst involves middle acidic sites which is consistent with the data obtained in works [2, 7]. Therefore, the promoting effect of polyvalent cations comes to the growth of such middle acidic sites.

Results of investigation of Pd-zeolite catalysts based on bicationic forms are shown in Figure 1-3.

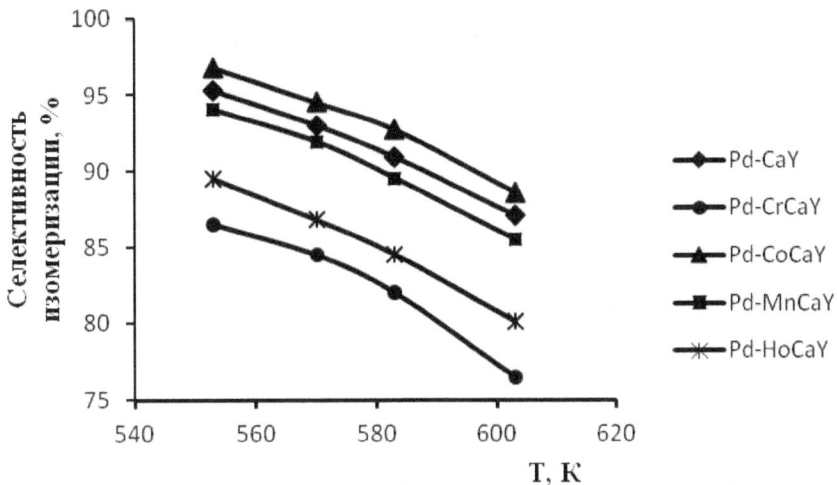

FIG. 1. Isomerization selectivity vs. temperature on Pd-MeCaY catalysts
Селективность изомеризации, % — Isomerization selectivity, %

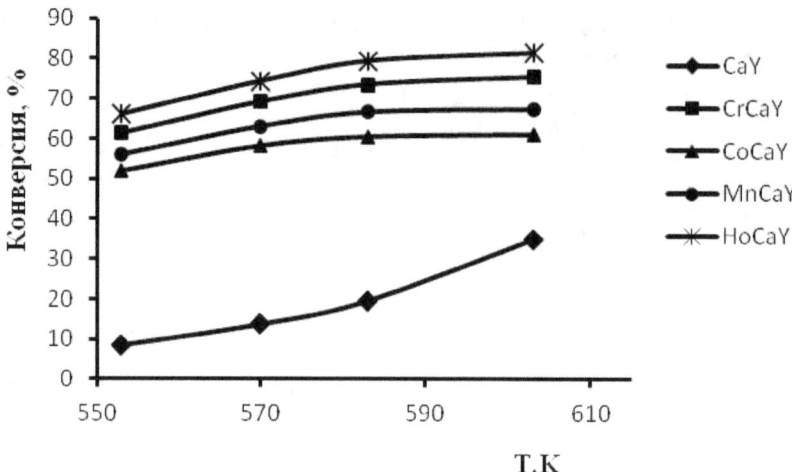

FIG. 2. Conversion vs. temperature on Pd-MeCaY catalysts
Конверсия, % — Conversion, %

Figure 1 demonstrates that Ho^{+3}, Mn^{+2}, Co^{+2} and Cr^{+3} cations provide promoting effect on Pd/CaY catalyst activity in n-heptane isomerization. Replacing 5% Ca^{+2} cations with Ho^{+3}, Mn^{+2}, Co^{+2} and Cr^{+3} dramatically increases Pd/CaY catalyst activity. Thus, if maximum isoheptanes yield over Pd/CaY catalyst is 32.7% (m/m), than isoheptanes yield over bicationic Pd-zeolite catalysts is 55.4–64.8% (m/m). Notice that the most promoting effect of Ho^{+3}, Mn^{+2}, Co^{+2} and Cr^{+3} cations is observed in n-heptane isomerization at low temperatures. Thus, at 570K, the difference in isoheptanes yield between Pd/MeCaY and Pd/CaY

catalysts is 40.4–50.1% (m/m), while at 603K it is 20.8–32.4% (m/m).

By comparing the activities of Pd-zeolite catalysts (Fig. 2 and 3) based on combination of calcium cations with Ho^{+3}, Mn^{+2}, Co^{+2} and Cr^{+3} cations, it was established that the catalyst based on combination of Ca^{+2} with Ho^{+3} is the most active in n-heptane isomerization. 64.8% (m/m) isoheptanes can be obtained over this catalyst. Among the investigated catalysts, the catalyst based on combination of Ca^{+2} and Co^{+2} cations shows low activity. Maximum yield over this catalyst of 55.4% (m/m) is achieved at 583K.

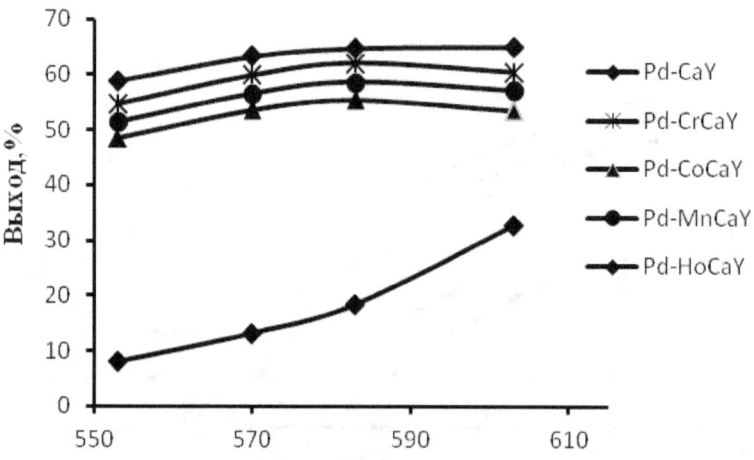

FIG. 3. Isoheptanes yield vs. temperature on Pd-MeCaY catalysts
Выход, % — Yield, %

The comparison of isomerazing selectivity shows that Pd-zeolite catalyst based on combination of Ca^{+2} и Co^{+2} has the most isomerazing selectivity (88.3-96.0% (m/m)) over the investigated range of temperatures 553–603K. Over the more active Pd-zeolite catalyst, high isomerazing selectivity (89.5-86.6) is only achieved in isomerization at low temperatures (553-570 K).

References

1. Kh.M. Minachev, Ya.I. Isakov. Metal-containg zeolites in catalysis. Nauka. — 1976. — 112 p.
2. S.E. Mamedov, B.A. Dadashev. Isomerization of n-hexane over Pd-zeolite catalysts containing rare-earth and transition elements. // Kinetika and kataliz. —1985. —V.36. — №1. — P. 236-238
3. J. Weitkamp, S. Ernst, C.Y. Chen. A Useful Catalytic Method for Probing the Effective Pore with of Molecular Sieves. // Stidies in Surface Science and Catalysis. — In. Proc. 8-th Intern. Zeolite Conf. Amsterdam. —1989. — V. 49. — №2. — P.2391-2396.
4. A.N. Vasilyev, P.N. Galich. Izomerization of n-paraffinic hydrocarbons over zeolitic catalysts. // Chemistry and Technology of Fuels and oils. — 1996. — V.32. — №4. — P 217-226.
5. W. Zhang, P.G. Smiriotis. Effect of zeolite structure and acidity on the product selectivity and reaction mexanizm for n-octane hydroisomerization and hydrocracking. // J. Catal. — 1998. —V.182. — P.400-416.
6. E. Blomsma, J. Martens, P. A. Jacobs. Izomerization and hydrocraking of heptane over bimetallic bifunctional PtPd/H-beta and PtPd/USY zeolite catalysts. // J. Catal. — 1997. —V.165. — P. 241-248.
7. J.F.M. Denayer, B.D. Ionckheere, G.B. Marin, G. Vanbusel, J.A. Martens Molecular competition of C_7 and C_9 n-alkanes in vapor- and liquid-phase hidroconversion over bifunctional Pt-USY zeolite catalysts. // J. Catal. — V. 2010. — P. 445-452.
8. Yushchenko V.V. Calculation of the catalysts acidity spectra based on the temperature-programmed adsorption of ammonia.// Journal of Physical Chemistry. — 1997. — V. 71. — No 4. — P.28-32.

Ozone Cross-Section Model and Its Effect on TOC Values Calculated from Spectral Irradiance Data

Vadim Stankevich

National Ozone Monitoring Center of Belarusian State University, Minsk, Belarus

Abstract. *Different ways to increase the accuracy of total ozone column retrieval from spectral irradiance data are discussed. Ozone cross-section model of Libradtran library is investigated as an accuracy increasing factor. Most suitable models for retrieving precise TOC values are found.*

Keywords: *ozone, total ozone column, spectral irradiance, libradtran.*

Ozone is the essential gas for the whole biosphere. It protects life on the Earth from hard ultraviolet (UV) radiation coming from the Sun. Ozone is also a greenhouse gas of high activity, so it takes significant part in climatic processes [1]. Thus climatic models created for long-term predictions of climate also must make predictions of total ozone column (TOC).

To build long-term predictions models, one should use long-term continuous observation data [2]. Using this data, one can find out trends in ozone distribution over geographic coordinates. Unfortunately, for the majority coordinates it is impossible because most commonly used direct-Sun (DS) methods of TOC retrieval work only when the sky is totally clear. Also they are unable to give precise data on low Sun zenith angles (SZA) [3]. To solve this problem, spectral irradiance data was proposed to be source for TOC values. Spectral irradiance data can be measured precisely even in bad measurement conditions.

Classical method of retrieving TOC from spectral irradiance data was proposed by Stamnes et al. In according to this method, one should use pre-calculated tables (Stamnes tables) that shows how TOC depends on SZA and spectral irradiance near the Earth surface [3]. Unfortunately, using of these tables does not imply the factors weaken the solar radiation in the UV range (atmospheric aerosols, clouds, etc.), which leads to considerable mistake in the results. It was estimated that by using Stamnes method the error of the TOC obtained values under unfavorable observation conditions approximates up to 30 %. On average, the error of the method is about 5 % [4]. This makes the use of this method for precision measurements virtually impossible.

It was shown before that instead of using Stamnes tables the inclusion of the factors weakening the solar ultraviolet radiation in the atmosphere (primarily aerosols influence) becomes possible by using optimization method based on calculating the minimum of the following function:

$$F(X, \alpha, \beta) = \sum_{X, \alpha, \beta} (R_{calc}(\lambda_1, \lambda_2, X, \alpha, \beta) - R_{obs}(\lambda_1, \lambda_2, X, \alpha, \beta))^2 \ (1),$$

where X — the required TOC; α and β — factors in the Angstrom's formula for atmospheric aerosol; R_{calc} — calculated spectral irradiance ratio of the atmospheric model for wavelengths λ_1 and λ_2; R_{obs} — retrieved in the issue of observations spectral irradiance ratio for the same wavelength.

Minimum of the function is at the point that corresponds to the actual values of TOC and to Angstrom's atmospheric aerosol parameters. For its acquisition Hook-Jeeves optimization pattern search is used. The calculated values of TOC are obtained using atmospheric models implemented in the library LibRadTran.

An optimization method is more accurate when using two pairs of wavelengths (see Fig. 2 and 3). In general, on clear days the average TOC values obtained using the optimization method deviate from satellite data for no more than 14 % for a couple of waves lengths and 8 % for the two couples [5].

It was also shown that while using optimization method, enumeration of nonselective adsorption of solar radiation by clouds makes results of TOC calculation more accurate. For this purpose, when measuring the Earth's surface spectral irradiation with spectroradiometer measurements are made not only in the UV range but also in the visible spectrum, after which the correction of changes in the UV signal channel is being made in accordance with a change in the visible channel signal

during the time of one measurement. The measurement accuracy can increase by 3 % on cloudy days and by 0.5 % on clear days with such an adjustment [6].

Further research of opportunities for increasing the accuracy of the TOC values by Earth's surface spectral irradiance in the UV range is associated with the choice of the optimal model for the calculation of the ozone cross-section implemented in the library LibRadTran. At present, there are three main models of ozone cross-section proposed by different authors: Bass and Paur (1985), Molina and Molina (1986) and Malicet et al. (1995) [7].

Calculations show that the data obtained by calculating TOC using a variety of ozone cross-section models differ in 1 % on average. TOC values calculated using Bass and Paur and Malicet et al. models correlate with each other and with the TOC values obtained from satellite data, more than TOC values calculated using Molina and Molina model, used in LibRadTran as a default one. The data obtained for one of the research days (10.04.2010) are shown in Figures 1–3.

In the result of ten days study, the average TOC deviation (calculated by Earth's surface spectral irradiance by solar UV radiation) from satellite data implies the following ranges: Bass and Paur model — 8,7 %; Malicet et al. model — 8.8 %; Molina and

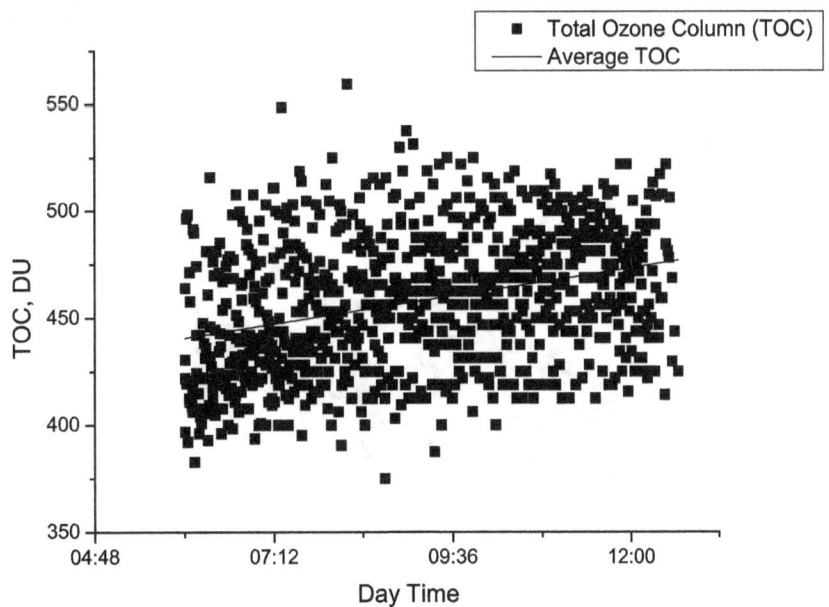

FIG. 1. TOC values obtained using the Bass and Paur model

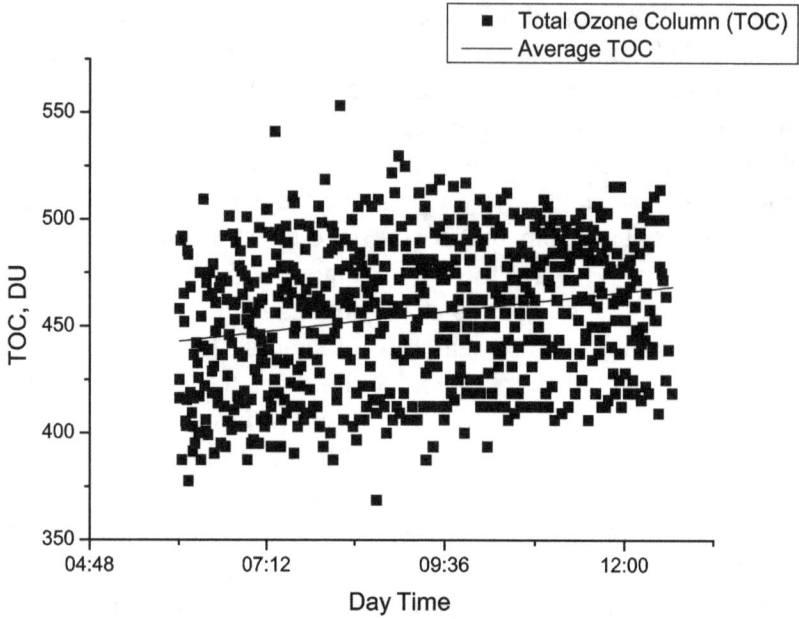

FIG. 2. TOC values obtained using the Malicet et al model

Molina model — 14.5 %. Data of the Earth's surface spectral irradiance by solar UV radiation were received in Minsk in 2010 with spectroradiometer «PION-UV», that runs measuring service at Minsk station of world ozonometric network.

Thus, to improve the accuracy of TOC values obtained using optimization method based on the Earth's surface spectral irradiation by solar UV radiation, it is advisable to apply the model for the ozone calculation developed by Bass and Paur and Malicet et al.

References

1. Petzoldt K. B., Naujokat, Neugebohren. Correlation between stratospheric temperature, total ozone, and tropospheric weather systems. // Geophysics Research Lett., 1994, v 21., p 1203 –1206.

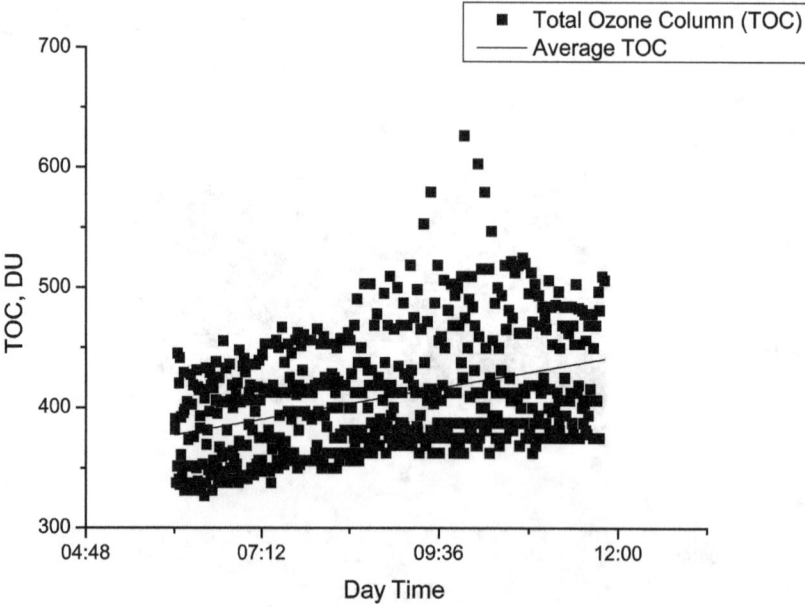

FIG. 3. TOC values obtained using the Molina and Molina model

2. Vanichec K., UV Index for public. // An Output from the COST-713 Project., Proceeding of the Quadrennial ozone symposium, Sapporo 2000, p.695–696.

3. Stamnes K. et al., Derivation of total ozone abundance and cloud effects from spectral irra-diance measurements. // Applied Optics, 1991, Vol. 30, No. 30, p. 1–15.

4. Svetashev A. G., Atrashevskij Ju.I., Krasovskij A. N., Stankevich V.Ju.. Predvaritel'nye rezul'taty vosstanovlenija OSO po rezul'tatam nazemnyh izmerenij spektral'nogo raspredelenija osveshhennosti UF sostavljajushhej solnechnoj radiacii. // Saharovskie chtenija 2009: Jekologicheskie problemy XXI veka. Materialy 9-oj Mezhdunarodnoj nauchnoj konferencii, Minsk, 21–22 maja 2009 g.,- s.305.

5. Stankevich V.Ju. Optimizacionnyj metod vosstanovlenija znachenij obshhego so-derzhanija ozona po spektral'noj plotnosti jenergeticheskoj osveshhennosti po-verhnosti Zemli solnechnym ul'trafioletovym izlucheniem. // Sostav atmosfery. Atmosfernoe jelek-trichestvo. Klimaticheskie processy. XV vserossijskaja shkola-konferencija molodyh uche-nyh, Borok, 30 maja — 4 ijunja 2011 goda, Tezisy dokladov, s. 53–54.

6. Stankevich V.Ju., Svetashev A. G., «Vlijanie oslabljajushhih faktorov na vosstanovlenie znachenij obshhego soderzhanija ozona v atmosfere». Vestnik BGU. Serija 1. № 2 / 2013.

7. Mayer, B. and Kylling, A.: Technical note: The libRadtran software package for radiative transfer calculations — description and examples of use. // Atmos. Chem. Phys., 5, 1855–1877, doi:10.5194/acp-5-1855-2005, 2005.

EARTH SCIENCE

Characteristics of Immobilized β-fructofuranosidase from *Saccharomyces vini* in Watery Medium

Davron B. Dekhkonov
Dilmurod Makhmudjanov
Namangan State University, Namangan, Uzbekistan

It's known that yeast enzymes have been using successfully in broad spectrum. Especially, yeast beta-fructofuranosidase seems to be more perspective for industrial purposes. The enzyme can be applied in bakery and confectionary to obtain invert syrup, in beverage technology to convert fusel oils into alkylfructosides [1]. It is important to investigate optimal enzymatic reaction conditions of enzymes before installation of enzyme preparations in scale up processes.

According to above mentioned problem, in the current work it was presented immobilized yeast beta-fructofuranosidase behavior in watery medium.

Materials and methods

Enzymes, chemicals and instruments. β-fructofuranosidase obtained from *Saccharomyces vini Rkaciteli-6* strain [2]. Glucooxydase (Reachem, Russia), Glutaraldehyde (MERCK, Germany), Silufol UV (Czechoslovakia), sucrose, D-glucose, dimethylformamide, urea, ammonium sulfate, n-butanol, acetic acid, isoamylol [Reachem, Russia] and other chemicals purchased from local sources (Chemreactivecomplect, Uzbekistan) were analytical pure. Centrifuge SLR-1 UCh-2, magnetic mixer MM5, ultra thermostats MTAKUTESZ TYPE-57 and auto-clave "Bergius-1L" used for the reactivation of activated carbon.

Hydrolytic activity of enzyme was determined by G.Afanaseva [3]. The process carried out in two steps. In first step 0,2 mkg (50 mkl) enzyme, 100 mkl 0,1 M acetate buffer pH 4,0 and 50 mkl sucrose was mixed and mixture incubated for 10 min at 50^0C. Reaction was stopped by adding 300 mkl 0,2 M K_2HPO_4 and enzyme inhibited in water bath for 5 min. As a control sample used inactivated native β-fructofuranosidase. For the immobilized β-fructofuranosidase used 8 mg immobilized preparation in place of 0,2 mkg (50 mkl) native enzyme. One unit of enzyme activity is defined as the amount of the enzyme that catalyses the hydrolysis of 1 Mm sucrose into glucose in 1 minute at 50^0C, pH 4,0.

Determination transferase activity of β-fructofuranosidase was defined using thin layer chromatography [4]. Solution containing 10–15 % sucrose and 25–40 % ethanol is divided two portions. First portion with 1–2 ml enzyme (0,4 mg/ml) and the second (control sample) is the same amount of inactivated enzyme in water bath. As a solvent used n-butanol, acetic acid and water (5:4:1) mixture. Silufol support transferred to solvent twice for the best migration and separation of alkylfructosides, fructose and sucrose. 5 gr urea, 20 ml HCl (2 N) and 100ml ethanol spray was used for appearance

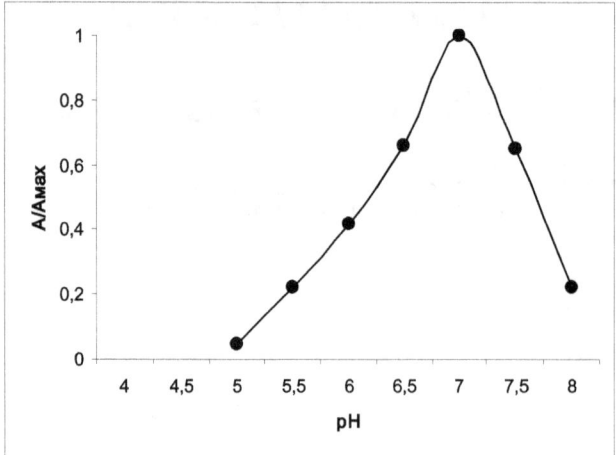

FIG. 1. pH optimum of immobilized β-fructofuranosidase

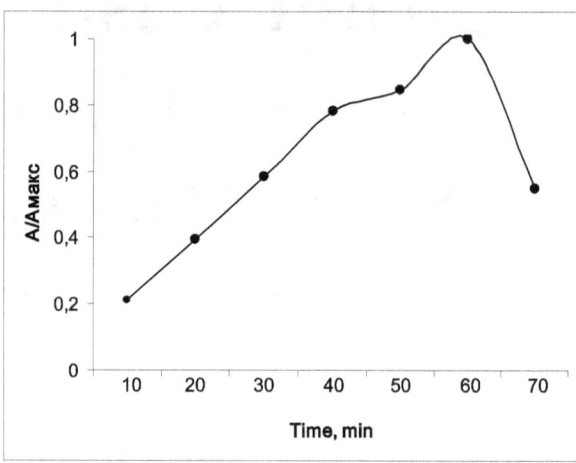

FIG. 3. Optimum of incubation time of immobilized β-fructofuranosidase

of formed products. Solid support was dried in thermostat at 105°C for 5 min.

Results

Determining of pH optimum studied in the range from 4,0 to 7,5 and the optimal found at 6,0. (Fig.1). 0,5 M sucrose in 0,1 M acetate buffer used as substrate. Increase of the optimal to alkaline condition can be explained by amino groups that were introduced into activated carbon.

Temperature optimum of the enzyme was performed between 20–70°C using 0,5M sucrose in 0,1M acetate buffer (pH 6,0). It was observed that the optimum of the enzyme laid at 30°C. (Fig. 2).

It was investigated influence of formed product (glucose) to the enzyme activity during established time. Due to result enzyme exhibited its maximal activity after 60 min and quantity of glucose employed 600 mkg in reaction mixture. After 70 min enzyme lost 44 % of global activity (Fig. 3).

Optimal concentration of substrate carried out 8–625mM intervals and the optimum of β-fructofuranosidase observed at 62,5mM (Fig.4). Reduction of enzyme activity in low substrate condition could be explained by law of mass action. Inhibition of β-fructofuranosidase in high concentration of sucrose (above 625mM) induced by substrate inhibition.

Dependence of immobilized β-fructofuranosidase activity and its quantity in reaction mixture

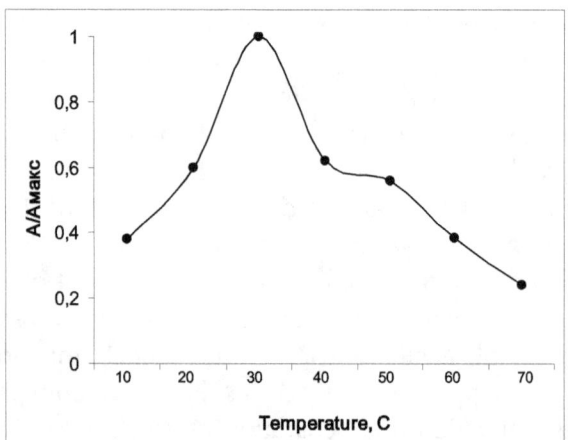

FIG. 2. Temperature optimum of immobilized β-fructofuranosidase

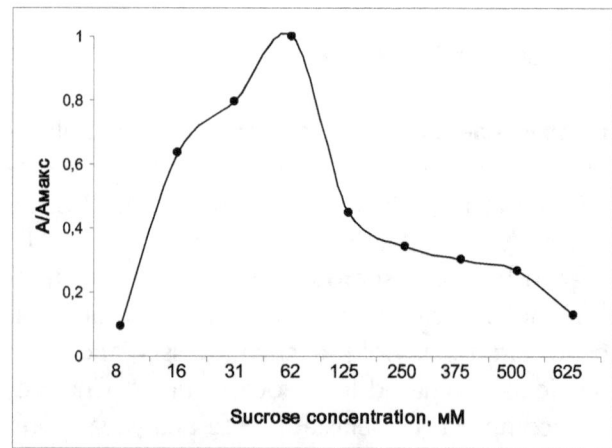

FIG. 4. Optimal concentration sucrose of immobilized β-fructofuranosidase

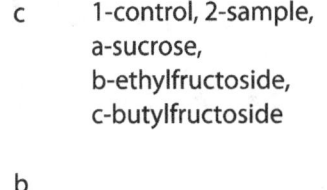

c   1-control, 2-sample,
     a-sucrose,
     b-ethylfructoside,
     c-butylfructoside

b

a

1   2

FIG. 5. Transferase activity of immobilized β-fructofuranosidase

was studied in the range of 8–108 mg of preparation that is responsible for 80–640 mkg of β-fructofuranosidase precoupled into sorbent. It was found that optimal quantity of the preparation employed 108 mg and it can be considered 640 mkg precoupled

β-fructofuranosidase. Transferase function of the enzyme was carried out by n-butylfructoside synthesis capability of β-fructofuranosidase in ethanol condition using n-butanol (Fig.5). Selected sample (n-butanol) was dropped 20 times on Silufol. The result of TLC showed that enzyme presented transferase activity after 12h. Using of the capability of immobilized β-fructofuranosidase, recommending preparation can be applied in beverage industry for the enzymatic transformation of fusel oils into alkyfructosides.

References

1. Abdurazakova S. H. Modernization of beverage technology using stimulated biocatalytic processes. Tashkent. FAN. 1990. -P.140
2. Dekhkonov D. B., Mirzarakhmetova D. T., Rakhimov M. M., Akhmedova Z. R. Obtaining high actively yeast β-fructofuranosidase. "Acta NUUz". Tashkent. 2004. Vol. 4. -P. 3–4
3. Afanaseva G. A., Sherbukin V. D., Modification of ferrocyanide variant of glucoxydase method for the determination glucose. Applied Biochemistry and Microbiology. 1975. Vol. 3. -P. 460–462
4. Oparin A. I. Nature and mechanism of action of yeast β-fructofuranosidase. M: Acad. Sc. SU 1955. -P. 22

Systematic Analysis of Dinophyta and Chrysophyta in Algoflora of Andijan Reservoir

Xilola Ergasheva

Namangan State University, Namangan, Uzbekistan

The location, climate, and geology of reservoirs are the most important factors in the evaluation and spread of water grasses. Andijan Reservoir is situated in Kyrgyzstan (Osh province) and Uzbekistan (Andijan region), 10 kilometers from the town of Xonobod and 70 kilometers from the city of Andijan. This reservoir was constructed on flat artificially created ground on the eastern side of Fergana Valley, around 700–800 meters above sea level. Andijan Reservoir is a hydrotechnical project which was built in the Kara Darya riverbed. Construction was completed in 1983and was based on planned irrigation-energetics approaches. The dam that forms Andijan Reservoir is 121 meters in height and 1,040 meters in length. When emergency flooding reaches 1700 m³/c, it is anticipated that water conductors in the dam will drain off excess water through irrigation channels that create water expenditure of 230 m³/c. The total capacity of the reservoir is 190 mln m³, and the effective capacity is 175 mln m³. The water level is 56 km², with an average depth of 6–8 m; the front of the dam is 25 m. The reservoir is 26 km long; the widest point is 12 km, the narrowest 1.5–2 km.

Construction of the reservoir began in 1965 and continued until 1983; the water has been in use since 1978. The climate of Andijan Reservoir is consistenly continental; summers are hot, cloudless and dry. According to information from the Scientific Research Hydrometerological Institute in Uzbekistan, the average temperature in July is +27.3 °C, and the average temperature in January is 3°C

(2006–2008). The water at the edge of the reservoir is covered with thin ice from December to the middle of February. In this territory the wind blows 15–20 m/sec, sometimes reaching 25–35 m/sec. The annual precipitation is 175–250 mm. The reservoir takes in water from the Kara Darya, Yassi, Tar, Karakulja, and Kurshab rivers. It is sierozem soil. The base of the dam stretches 10–15 m on the left side to the cliffs, 8–12 m on the right side. The reservoir can provide water for crop fields of 266,700 hectares and additionally irrigate 32,800 hectares.

A systematic analysis of component species of algoflora in Andijan Reservoir showed a total of 418 species and species sorts, water grasses that are due to forms. There are 6 branches, 12 classes, 20 orders, 49 families, 116 series, 400 species, 9 forms and 9 species sorts (Table 1).

According to determined branches of algoflora component species, Table 1 shows 107 species of *Cyanophyta* branch water grasses; 5 species of *Chrysophyta* branch; 23 species of *Euglenophyta* branch; and 151 species of *Chlorophyta* branch. Our analysis showed that comparatively few of the water grasses species belong to *Dinophyta* and *Chrysophyta* branches. In this branch water-grasses were analysed with tacsonomics. Water grasses of *Dinophyta* branch totaled 17 species (4.0 %); they are 1 class, 1 order, 1 family, and 3 series. *Dinophyceae* class, *Peridiniales* order, *Peridiniaceae Pauls* family, *Glenodinium (Ehr.)* Stein. Series 8 species, *Peridinium Ehr.* series 5 species, *Ceratium Schrauk* series 4 species and species sorts were found. From

TABLE 1. Systematic Analysis of Algoflora in Andijan Reservoir (2006–2009)

| Branches | Number of Tacsons | | | | | Species Sorts | |
	Class	Order	Family	Series	Species	Variations	Forms
Cyanophyta	2	4	16	28	100		7
Chrysophyta	2	2	2	2	4	1	
Bacillariophyta	2	4	8	29	112	3	
Dinophyta	1	1	1	3	15		2
Euglenophyta	1	1	1	5	23		
Chlorophyta	4	8	21	49	146	5	
Total:	12	20	49	116	400	9	9

this branch *Glenodinium (Ehr.)* Stein series leads in algoflora with numerous species. Systematic analysis of *Dinophyta* branch is given in Table 2.

Systematic analysis of *Chrysophyta* branch water grasses found 5 species(1.1 %), they are 2 classes, 2 orders, 2 families, and 2 series.

Chrysomonadineae class, *Ochromonadale* order, 4 species series owing to *Euochromonadaceae Pasch;* and *Chrysocapsineae* class, *Hydrurales* order, *Hydruraceae Pasch.* family from *Hydrurus Kirchn* series one species *H. foetidus Kirch* determined.

The composition of algoflora branches of water grasses in Andijan Reservoir is as follows: *Chlorophyta* (36.1 %), *Bacillariophyta* (27.7 %), *Cyanophyta* (25.6 %), *Euglenophyta* (5.5 %), and *Chrysophyta* and *Dinophyta* together (5.3%).

In algaflora components of Andijan Reservoir, there are particular reasons why *Dinophyta* and *Chrysophyta* branches occur less often. First, water grasses from these two branches show no adaptation to the ecological conditions of the environment. Second, their growing activity stops during the hottest and coldest months of the year. Third, they spread more in fast-flowing water basins than in reservoirs. Fourth, they exist separately with no contact with other branch species. Besides that, worldwide there are 42,110 species of water grasses. By comparison, *Dinophyta* at 500 species and *Chrysophyta* at 400 species are much fewer in number. Our analysis has confirmed that *Dinophyta* and *Chrisophyta* water grasses occur less frequently than other water grasses in Andijan Reservoir.

TABLE 2. Systematic Analysis of Water Grasses in *Dinophyta* branch

| Systematic Singulars and Their Number | | | | |
Class	Order	Family	Series	Species and species sorts number
Dinophyceae	Peridiniales	Peridiniaceae	Glenodinium Ehr.	8
			Peridinium Ehr.	5
			Ceratium Schrauk	4
Total:	1	1	3	17

TABLE 3. Systematic Analysis of Water Grasses in *Chrysophyta* branch

Systematic Singulars and Their Number				
Class	Order	Family	Species	Species and species sorts number
Chrysomonadi Neae	Ochromona-dales	Euochromonada-ceae Pasch	Dinobryon Her	4
Chrysocapsi Neae	Hydrura-Les	Hydruraceae Pasch	Hydrurus Kirchn	1
Total: 2	2	2	2	5

References

1. Starmach K. Cyanophyta, Polska Akad. Nauk. — Warszawa, 1966. — 216 p.
2. Utermohl H. Zur Vervollkommnung der quantitativen Phytoplankton- Methodik // Verein. Limnol. — Mitteilungen, 1958. — № 9. — P. 1–38.
3. Wislouch S. M. Beitrage zur Diatomeen flora von Asien, L, Die Diatomeen des. Balchosch // Sees. Ber. d. d. Bot. Gesellsch., 1956. — № 41 (8). — P 17–21.
4. Starmach K. Cyanophyta, Polska Akad. Nauk. — Warszawa, 1966. — 216 p.

Ecological and Epidemiological Situation of Toxocariasis in Russia

Viktoria Erofeeva[1]
Olga Maslennikova[2]
Vera Puhlyanko[1]

[1]Peoples' Friendship University of Russia, Moscow, Russia
[2]Vyatka State Agricultural Academy, Kirov, Russia

Abstract. *The article describes the epidemiological situation of toxocariasis in Russia, transplacental transmission of larvae cysts to newborn, Toxocara canis infestation of final hosts. Data on soil infestation with toxocara eggs are summarized. We have detected an unusual localization of mature T.canis female in the dog's lung. Canines form natural synanthropic foci, biological activity of which complicates the epizootological situation of toxocariasis in Russia.*

Keywords: *epidemiology, human, toxocariasis, dogs.*

Toxocariasis is a parasitic disease caused by the migration of the larvae of dog helminths - *Toxocara canis*, rarely - of cat helminths - *Toxocara mystax* in the human body and characterized by complex syndromes and symptoms, referred to as *visceral larva migrans*.

The incidence of toxocara in different countries varies from 2.6 % in Belgium to 80 % in the Caribbean. In Russia, according to research carried out in different regions, toxocara infested half a million people. Annual growth of the incidence of toxocariasis is observed, especially in rural areas among children. In case of massive invasion of toxocara, severe diseases, primarily, of respiratory organs may develop in the human immune system. This highlights the **relevance of the problem** of toxocariasis for the assessment of environmental and epidemiological well-being of regions.

Toxocariasis is a disease relatively little known to practitioners. However, toxocara infestation is widespread among both animals and humans. According to WHO (1980), infestation of canines - the main hosts of toxocara - is very high worldwide, constituting 90 % in some regions. Increase in the number of dogs in the cities, high rate of toxocara infestation of them, intensity of egg excretion by mature helminths inhabiting the intestines of animals, eggs stability in the external environment, are crucial factors in the spread of infestation among the people. Toxocara infestation of cats and their role in spreading the infestation among humans is underexplored.

Toxocariasis is a zoonotic infestation for human. It is characterized by severe, prolonged and recurrent course, polymorphism of clinical manifestations caused by the migration of toxocara larvae to various organs and tissues. Human is infected by swallowing infective toxocara eggs [5].

Epidemiology: A clearly pronounced upward trend in detection of toxocariasis is marked in Kirov region, as in the whole of Russia due to wide introduction of the techniques of its diagnostics into health care practice. If in 1997 Kirov region had only one case of toxocariasis in human registered, in 2002 this figure constituted already 19. The incidence of toxocariasis increased by 2 times and was 2.3 per 100 thousand people in 2006 compared to 2005 (according to the Federal Service on Surveillance for Consumer Rights Protection and Human Well-being in Kirov).

According to V.P. Sergiev et al. [12], the prevalence of toxocariasis significantly exceeds officially reported figures due to its dynamism and contingence with somatic pathology. The number of toxocariasis cases has increased by 80 times in 2003 compared to 1991. According to the Federal Service on Surveillance for Consumer Rights Protection and Human Well-being, toxocariasis ranks sixth in the RF parasitosis structure. The incidence of this helminthiasis has increased by 2 times in 2009 compared to 2001 and constituted 1.6 per 100 thousand people in adults and 7.7 - in children according to official statistics. In 2011, the total toxocariasis infestation was 2.32 per 100 thousand people.

Toxocariasis is a serious problem in Kurgan region. In 2012, the incidence of toxocariasis has increased by 57.3 % compared to 2010 and by 12.2% compared with 2011. In 2012, 361 cases of toxocariasis were registered in total (39.72 per 100 thousand people). The incidence of toxocariasis in children up to 17 years has increased by 2.5 times compared to 2010 and by 1.2 times compared to 2011. In 2012, 199 cases of toxocariasis among children under 17 were registered -117.1 per 100 thousand people. [4]

The source of human invasion in synanthropic focus is dogs contaminating soil by the eggs of toxocara, excreted in faeces. Man is a deadlock for toxocariasis pathogen. People infected with toxocara cannot be a source of infestation because parasite does not get maturity in human body and does not excrete eggs into the external environment. Human serves as a reservoir or parathenic host for toxocara and can actually be construed as a "deadlock" of toxocara pathogen [14]. Dogs are infested with toxocara in several ways: intrauterine infection of puppies through the placenta of infested pregnant lady-dog, through milk of feeding dog (transmammary way), ingestion of infective eggs by dogs from the soil contaminated with toxocara eggs, ingestion of puppies' faeces infested with toxocara by adult dog in the lactation cycle, ingestion by dogs of tissues of parathenic (reservoir) hosts with infective larvae [1,3,10,16].

Basically, the source of human invasion is dogs, and also foxes and wolves in the countryside. Unusual localization of mature *Toxocara canis* female was detected in 4-month-old puppy dog in the left lung bronchus which died from the plague. Its intestine had no toxocara, because dehelminthization was conducted earlier (1.5 months ago), following which a large number of *T. canis* died. When puppy's lung was opened, spirally twisted female of *T. canis* jumped as a spring from the left bronchus. The length of female was 10.6 cm [8].

Newborn wolf and fox cubs are 100% infected with toxocara, according to our research. Adult wolves get rid of toxocara, but foxes are infected with them in adulthood as well. *T. canis* in adult lynx was recorded. [7] In recent years, according to our data, foxes of Northeast European Russia rarely have *Toxascaris leonine*, it was replaced by *Toxocara canis*.

However, direct contact with canines does not play a key role in the contagion process, as toxocara eggs excreted by dogs are non-invasive and require maturation in the external environment. Soil contaminated with feces of infected dogs, and wild berries and mushrooms in the countryside play major role in the transmission of invasion to human. Toxocara eggs can also be transmitted with vegetables and herbs. It has been established that toxocara eggs are detected at up to 3 % of the number of wipe-samples from vegetables, berries and greenery grown in small holdings. Other transmittance factors include animal dander, contaminated food, water, hands [5].

The parasitologic assessment of 570 water samples in the urban area of Kursk region established that 151 samples (26.5%) contained Giardia cysts, Cryptosporidium oocysts, ascarid and toxocara eggs. The number of detected pathogens per sample was 6.2 specimens. Ascarid and toxocara eggs, Giardia cysts were found in drinking water (357 samples) - 2.9 specimens per sample. The results of parasitologic analysis of 1258 soil and sand samples showed parasitic pathogens in 511 (40.6 %) cases. Toxocara, ascarid and whipworm eggs were found in soil. Contamination of soil in recreation areas was 59.5 % and of sand in playgrounds of residential units - 49 %. On average, 3.5 to 8.6 pathogens were recorded per soil sample and 1.6 to 8.6 pathogens per sand sample. [6]. Soil contamination with toxocara eggs in Astrakhan region increases every year. Toxocara eggs were found in 6.6 % of samples in 2004, 10,8% in 2008 [11]. 402 soil samples from different areas underwent sanitary parasitological examination in Rostov region. *T. canis* eggs share of total number of parasites identified in all areas averaged 65.44 % with a range from 13.3% to 100% in different areas of the region. The intensity of soil contamination with eggs of this pathogen ranged from 1.2 to 16.67 specimens per 1 kg of soil. Egg viability ranged from 20 % to 84 % [15].

The risk group in terms of toxocariasis infection includes children aged 1.5-5 years in contact with the soil and dogs; children suffering from geophagy, persons having regular professional contact with animals and soil (veterinary workers, workers of dog kennels, circuses, zoos, workers of utilities, vendors of vegetable shops, workers of vegetable depots, etc.), mental defectives and the mentally infirm with a liability to coprophagy and geophagy; owners of farmlands; pet owners; persons engaged in hunting with dogs or other outside activities involving dogs.

Although the source of toxocara invasion for humans is basically dogs, direct contact with them does not play very special role in human contagion. It has been established that soil as a pathogen transmission factor ranks first in order of importance. Dogs excrete toxocara eggs in faeces into the environment and these eggs mature in the soil up to the infective stage. Eggs can remain viable in the soil throughout the year, overwintering under snow well in central Russia. The development of eggs to the infective stage requires 160-183 degree-days. It would take about 36 days at an average daily temperature of + 13-18 ° C and about 15 days at a temperature of +25 ° C. Toxocara eggs remain viable in the soil for several years [5].

Existence of several ways of toxocara pathogen spread in dogs is a cause of very high rate of

their infestation. The number of dogs in the world is huge and is constantly increasing. According to K.I. Skryabin All-Russian Institute of Helminthology, the number of dogs in Moscow (registered and stray) is more than 1 million subjects which leave to 270 tons of fecal masses daily on its territory. Recent studies of K.I. Skryabin All-Russian Institute of Helminthology, E.I. Martsinovskiy Institute of Medical Parasitology and Tropical Medicine, I.M. Sechenov Moscow Medical Academy showed that 42-46 % of fecal samples of dogs contained *Toxocara canis* eggs. The number of helminth eggs in 1 g of dog feces can reach 40 thousand [12, 13]. In later studies, toxocariasis was recorded in dogs in both urban and rural areas under the conditions of nonchernozem belt of Russia. Extensity of infestation, on average, accounted for 58.1% in urban dogs and 64.3 % in rural dogs. High infestation of dogs with *Toxocara canis* was established in all seasons and especially during the summer- autumn period. The maximum number of toxocara eggs per 1 g of dog feces was recorded in summer and in September - up 220.6 specimens. [2].

With such a large number of dogs and given that many of them are stray, the problem of environmental contamination with dog faeces is becoming more acute. This is also explained by the limited number of pet relief areas and poor sanitary knowledge of dog owners.

Toxocara transmission paths include direct path and with the involvement of an optional host (one of options): **definitive host (canines)** - soil - optional (reservoir, parathenic) host - **definitive host (canines)**. Invasion transmission mechanism in this option is geooral - xenotropic. Rodents, pigs, sheep, birds, earthworms can be parathenic (reservoir) host.

The pathogen is developed further, provided that the parathenic host was eaten by dog or other definitive host. Human is the parathenic host as well, but he is not included in the invasion transmission cycle, being a biological dead end for the parasite. Human is infected with toxocara through contact with the ground (playing in sand-pit, performing excavation works for construction, working in the garden, conducting repairs, etc.), contaminated with toxocara eggs from infested dogs. Less often, infection is possible through direct contact with dogs, paws and coat of which pick up soil containing mature toxocara eggs outdoors. There are observations of human infection through eating raw liver as well as other raw or poorly thermally processed meat: lamb, rabbit, chicken [14].

Recently, there is evidence of the possible transplacental transmission of toxocariasis in humans. The available literature contains only two reports of toxocariasis in newborn as a result of transplacental transmission of larvae cysts from the mother diseased during the pregnancy. One case was registered in Egypt (Oteifa at all., 1996), the second - in Russia, Orenburg. Congenital toxocariasis was identified through serological studies. Child from the first pregnancy, born 01.02.2000, the mother at the age of 21. Immediately after birth, when mucus was sucked out of the nasal passages - hemorrhagic secretion. When mucus was sucked out of the stomach - red blood, 1 hour after birth bleeding occurred again. The child had dyspnea, acrocyanosis, heart stones were muffled. The child was in serious condition. Bright pink erythematous foci were detected on the skin of the buttocks, feet, legs, on the background of which vesicular rash was observed, dull yellow secretion during the medical examination of 04.02.2000. A blood test revealed a high eosinophilia in a child (under 51 %) and mothers (31 %). The blood of the child and the mother was tested for anti-Toxocara antibodies. Antibody titers in the ELISA (enzyme-linked immunosorbent assay) constituted 1/800 in the mother, 1/400 in the child. The child was diagnosed with "congenital toxocariasis infection". The child was treated with dekaris for 5 days [9].

Prevention of toxocariasis is a state concern in Russia. Its addressing should involve executive authorities, housing management organizations, animal disease control stations, health authorities with the participation and under the control of sanitary and epidemiological authorities. Toxocariasis prevention efforts should be undertaken in several directions.

1. Activities aimed at the main sources of infestation. These include primarily examination and timely dehelminthization of dogs.
2. Impact on the invasion transmission factors.
3. Hygienic measures.

Since the majority of the population is not aware of the risk of worm infestation from dogs, educative activities among the population which should include an explanation of the possible ways of infestation, methods of animal treatment, the need to remove fecal contamination from dogs during their walking are important. To this end, it is useful to adopt the practice of some European countries, which have organized installation of special containers for plastic bags for dog faeces to be collected in places where dogs are walked most frequently for the protection of parks and gardens from fecal contamination of dogs [14].

Conclusion: Toxocariasis is relatively pressing environmental and epidemiologic problem. Its solving depends to a large extent on focused collaborative work of medical and veterinary services, as well as on the introduction of the latest invasion diagnostics, treatment and prevention methods and techniques into the health care practice.

References

1. Alekseeva M.I. Toxocariasis: clinical features, diagnostics, treatment // Med. parasitology and parasitic diseases. - 1984. – No. 6. - P. 66 - 72.
2. Arkhipov I.A., Zubov A.V., Borzunov E.N., Mikhin A.G. Veterinary and sanitary problems of parasitology caused by the increased number of dogs and cats in cities // Theory and practice of parasitic disease control: Materials of scientific conference report, Moscow, 2009. – No. 10 - P. 22 - 26.
3. Velichkin P.A., Radun F.L. Experience of recovery from toxocariasis of foxes and Arctic foxes of Saltykov state fur farm and dogs of service dog kennel // Helminthosis in humans, animals, plants and their control: Conference report thesis, All-Soviet Union helminthologists' community of the Academy of sciences of the USSR. - M., 1980. - P. 31 -33.
4. State report "On the state of sanitary and epidemiological welfare of the population in Kurgan region in 2012. Kurgan, 2013. - P. 85 - 86.
5. Lysenko A.Ya., Konstantinova T. N., Avdyukhina T.I. Toxocariasis: Textbook. Russian Medical Academy of Postgraduate Education. Moscow, 2004. – 40 p.
6. Malysheva N.S., Samofalova N.A., Plekhova N.A., Borzosekov A.N. Parasitologic assessment of habitat quality in urban areas of Kursk region // Theory and practice of parasitic disease control: Materials of scientific conference report, Moscow, 2009 . No. 10 - P. 255 - 257.

7. Maslennikova O.V. Helminthofauna of game animals in natural biocenosis of Kirov region // Synopsis of the thesis ... Cand. Sc. {Biology} - M., 2005 – 20 p.

8. Maslennikova O.V., Maslennikova T.V. Spread of Toxocara canis (Werner, 1782) in natural biocenosis of Kirov region and some features of its localization // Theory and practice of parasitic disease control: Materials of scientific conference report, Moscow, 2008. No. 9. - P. 289 - 292.

9. Nazarenko S.I., Mozgova L.A., Nazarenko L.V. Identified case of congenital toxocariasis // News-bulletin "Vektor –Best" - 2000, - No. 2 (16), June. - P. 16.

10. Petrov A.M. Helminthic diseases of fur animals. M.: International Book, 1941. - 226 p.

11. Postnova V.F., Shendo G.L., Dzharkenov A.F., Bazel'tseva L.I., Postnov A.B., Okunskaya E.I. Asessment of epidemiologic significance of soil at toxocarosis // Theory and practice of parasitic disease control: Materials of scientific conference report, Moscow, 2009. No. 10 - P. 304 - 306.

12. Sergiev V.P., Uspenskiy A.V., Romanenko N.A. et al. New and returning helminthiasis as a potential factor of the socio-epidemic complications in Russia // Medical Parasitology. - 2005. – No. 4. - P. 6 - 8.

13. Sonin M.D., Bessonov A.S., Roitman V.A., Sergiev V.P. Moscow metropolis environment and parasitic contamination problems // Medical Parasitology. 1995. – No. 3. P. 3 - 7.

14. Tumol'skaya N.I., Sergiev V.P., Lebedev M.N. et al. Toxocariasis. Clinical features. Diagnostics. Treatment. Prevention. - Moscow, 2004. - 48 p.

15. Shishkanova L.V., Vaserin Yu.I., Khromenkova E.P., Dimidova L.L., Upyrev A.V., Tverdokhlebova T.I., Prigodin A.V. Soil contamination with helminth eggs in Rostov region // Theory and practice of parasitic disease control: Materials of scientific conference report, Moscow, 2009. No. 10 - P. 439 – 441.

16. Sprent J. F. A. Observations on the development of Toxocara canis (Werner, 1782) in the dog //J. Parasitol. — 1958. — 48. — N 3. — P. 184–209.

Nonlinear Resonance Magnetoelectric Effect in Magnetostrictive Piezoelectric Structures

Tatyana O. Firsova

Yaroslav-the-Wise Novgorod State University, Velikiy Novgorod, Russia

Abstract. *The results of theoretical investigations of the nonlinear magnetoelectric effect in magnetostrictive piezoelectric structures have been presented. It has been shown that in such structures resonant excitation of the electric field occurs by means of an alternating magnetic field with a frequency that is half the frequency of the electromechanical resonance. In the low frequency range of the spectrum two voltage peaks arises due to the superposition of the signals from the linear and the nonlinear effects, the value of the voltage difference of two adjacent peaks voltage is proportional to the static magnetic field. The peak value of the electric field of the nonlinear magnetoelectric effect doesn't depend on the bias magnetic field.*

Keywords: *nonlinear magnetoelectric effect, magnetostriction, magnetic, piezoelectric.*

1. Introduction

Magnetoelectric (ME) effect consists in inducing of the polarization by an applied magnetic field (direct ME-effect) and, vice versa, in inducing the magnetization by an external electric field (inverse ME-effect). This effect is interesting because it allows to create fundamentally new solid state electronics devices [1, 2]. Recently, the linear ME effect was investigated in detail enough; however, there are only a few work in which the nonlinear ME-effect were observed. The origin of the ME effect in magnetostrictive piezoelectric structures is the mechanical interaction between magnetostrictive and piezoelectric phases. The direct and inverse linear ME effects in such structures were investi-

gated in details. But the magnetostriction is quadratic function on the magnetization [3] therefore the magnitude of mechanical stress in the region far from saturation is proportional to the square of the magnetic field. Therefore electric field on the capacitor will also be proportional to the square of the magnetic field. This leads to the fact that an electric voltage will be produced in an alternating magnetic field with a frequency ω on the capacitor plates with doubled frequency. If the applied magnetic field is the sum of two fields - alternating magnetic field and static magnetic field, then the electric voltage on the capacitor is the sum of two signals. First signal is the signal producing the linear effect in the alternating magnetic field with the frequency ω. Second signal is the nonlinear effect with double frequency. The signals sum with single and double frequency leads to the occurrence of the signal amplitude difference between the values of adjacent peaks voltage on the capacitor plates. The magnitude of this difference is proportional to the applied static magnetic field. This effect can be used to measure the static magnetic field. Therefore, devices can be created based on the magnetoelectric effect, such as magnetic field sensors, the sensitivity of which are significantly higher than the sensitivity of Hall sensors [4].

2. Model

As a model, we consider a structure in the form of a rectangular bar, consisting of a magnetic layer with thickness $^m t$ and a piezoelectric layer with thickness $^p t$. The piezoelectric layer is initially polarized perpendicular to the sample plane (along the Z axis). We assume that the directions of the static magnetic and alternating fields with frequency ω coincide with the direction of polarization. The alternating magnetic field induces elastic oscillations at the frequency ω in the magnetostrictive layer. The oscillations are transferred across the interface into the piezoelectric layer via a shear stress, which results in coupled oscillations of the subsystems. Assuming that the bar is narrow, the displacements can be regarded as homogeneous along the Y axis and the only nonzero components of the stress tensor are Txx and Txz. Since there is a sharp interface, across which the magnetostrictive and piezoelectric layers interact, the magnitude of the stress is inhomogeneous over the thickness of the sample, perpendicular to the interface. The equation of motion for x projection of the displacement vector of the medium, taking into account the heterogeneity of oscillation in a direction perpendicular to the interface is given by equation:

$$^\alpha \rho \frac{\partial^2 {}^\alpha u_x}{\partial t^2} = \frac{\partial {}^\alpha T_{xx}}{\partial x} + \frac{\partial {}^\alpha T_{xz}}{\partial z} \qquad (1)$$

where index α is equal to m for the magnetostrictive and p for the piezoelectric layer, respectively, $^\alpha \rho$ is density of ferrite or a piezoelectric material, $^\alpha T_{\bar{j}}$ is stress tensor.

Assuming that a plate is long and narrow, i.e. neglecting the width of the plate irregularities, the

equations for the components $^m S_x$, $^m S_z$, $^p S_x$, $^p S_z$ of the strain tensors and the z projection of the electric induction D_z of the polarized piezoelectric phase take the form:

$$^m S_x = \frac{1}{^m Y} {}^m T_x + g_{x,z}(H_z)^2, \qquad (2)$$

$$^m S_z = \frac{1}{^m G} {}^m T_z, \qquad (3)$$

$$^p S_{xx} = \frac{1}{^p Y} {}^p T_{xx} + {}^p d_{xx,z} {}^p E_z, \qquad (4)$$

$$^p S_z = \frac{1}{^p G} {}^p T_z. \qquad (5)$$

$$D_z = \varepsilon_z E_z + d_{z,x} {}^p T_x. \qquad (6)$$

where $^m Y, {}^m G, {}^p Y, {}^p G$ are <u>Young's modulus</u> and the shear modulus of the magnetostrictive and piezoelectric phases, respectively, $d_{z,x}$ is the piezoelectric tensor, ε_z – is the dielectric tensor, $H_z = H_m exp(i\omega' t)$ is the intensity of the alternating magnetic field with a frequency ω', E_z is z- component of the electric field.

The solution of equation (1) represented in the form:

$$^\alpha u_x = {}^\alpha u(x,z) \exp(i\omega t), \qquad (7)$$

where $\omega = 2\omega'$ is the frequency of mechanical vibrations.

Let us represent the solution of the equation for the displacement vector of the medium in terms of plane waves, whose amplitude varies across the thickness of the sample:

$$^\alpha u_x = {}^\alpha g(z) \cdot ({}^\alpha A \cos(\omega t - k) + {}^\alpha B \sin(\omega t - k), \qquad (8)$$

where $^\alpha A$ and $^\alpha B$ are the integration constants and $^\alpha g(z)$ is some function.

Substituting Eq. (8) into equations of motions (1) leads to the equation for the function $^\alpha g(z)$. After simple transformations, the equations for the functions, which determine the variation of the oscillation amplitude, take the form:

$$^m g''(z) + 2(1+v)\left[\frac{\omega^2}{^m V^2{}_L} - k^2\right] {}^m g(z) = 0 \qquad (9)$$

$$^p g''(z) + 2(1+v)\left[\frac{\omega^2}{^p V^2{}_L} - k^2\right] {}^p g(z) = 0 \qquad (10)$$

where we have introduced the notations $^mV_L=\sqrt{^mY/^m\rho}$, $^pV_L=\sqrt{^pY/^p\rho}$ for the velocities of longitudinal waves in the magnet and piezoelectric, respectively, and v is the Poisson's coefficient, which is assumed to be the same for both media. The form of the functions $^mg(z)$ and $^pg(z)$ (exponential or trigonometric) depends on the sign of the term in square brackets in Eqs. (9) and (10), which is in turn given by the relation between the speeds of sound in the magnet and piezoelectric. For certainty, we choose the most common case, in which the speed of elastic modes in the magnet is higher than in the piezoelectric.

$$^mg(z)=\tilde{N}_1\exp(^m\chi\,z)+\tilde{N}_2\exp(-^m\chi\,z),\quad(11)$$

$$^pg(z)=\tilde{N}_3\cos(^p\chi\,z)+C_4\sin(^p\chi\,z),\quad(12)$$

with the notations

$$^m\chi^2=-2(1+v)\left(\frac{\omega^2}{^mV_L^2}-k^2\right)$$

$$^p\chi^2=2(1+v)\left(\frac{\omega^2}{^pV_L^2}-k^2\right)$$

Using the open-circuit conditions we can get ME coefficient equation in the following form:

$$U(t)=\frac{^pY\,d_{x,z}\,g_{x,z}}{\mathring{a}_z\,\ddot{A}_a}\frac{^mY^mt}{^mY^mt\frac{h(^m\hat{e})}{^m\hat{e}}+^pY^pt\frac{g(^p\hat{e})}{^p\hat{e}}}$$
$$\frac{g(\hat{e})}{\hat{e}}\frac{g(^p\hat{e})}{^p\hat{e}}\,^pt\,(H_z(t))^2$$

(13)

where

$$\Delta_a=1-K_p^2\left(1-\frac{^pY^pt}{^mY^mt\frac{h(^m\kappa)}{^m\kappa}+^pY^pt\frac{g(^p\kappa)}{^p\kappa}}\frac{g(\kappa)}{\kappa}\frac{g(^p\kappa)}{^p\kappa}\right),$$

$$K_p^2=\frac{^pY(d_{x,z})^2}{\varepsilon_z}$$ is the square of the electromechanical coupling constant, $^m\kappa={}^m\chi^mt$ and $^p\kappa={}^p\chi^pt$ are the dimensionless variables, v is

Poisson's ratio, which is assumed to be the same for the magnetostrictive and piezoelectric phases, $\kappa=k/2$ is dimensionless parameter, k – wave number.

Equation (14) describes the frequency dependence of the ME effect. As can be seen from equation (14), at so-called antiresonance frequencies, when $\Delta a=0$, the peak increasing of the ME voltage coefficient is observed. It should be noted that the antiresonance frequency is near to the resonance frequency, which is determined by condition $L=(\lambda/2)(2n-1)$, where λ- the wavelength of acoustic waves, n=1, 2...- an integer number. Thereby the frequency of the first resonance is observed near the frequency determined by the expression:

$$f_{res}=\frac{1}{2L}\sqrt{\frac{^mY\,^mt+^pY\,^pt}{^m\tilde{n}\,^mt+^p\tilde{n}\,^pt}}\quad(14)$$

4. Conclusion

In magnetostrictive piezoelectric structures, along with the linear, there is also a nonlinear ME effect. This effect leads to a resonant excitation of the electric field under the influence of an alternating magnetic field with a frequency that is double lower than the electromechanical resonance frequency. The value of the nonlinear effect is quadratic in the alternating magnetic field and is independent of the bias field.

References

1. G.A. Smolensky, I.E. Chupis. Ferroelectromagnets Sov. Phys. Usp. 25 475 (1982).
2. Y.N. Venevtsev, V.V. Gagulin, V.N. Lyubimov Nauka, M., 224, (1982).
3. K.P. Belov Magnetostrictive phenomena and their technical application Nauka, M., 160, (1987).
4. J.Y. Zhai, Z.P. Xing; S.X. Dong, J.F. Li; D. Viehland Detection of pico-tesla magnetic fields using magneto-electric sensors at room temperature Appl. Phys. Lett. 88, 062510 (2006).

Technology Development of Wastewater Treatment for Climate of Uzbekistan

Mukhtor G. Khurramov

Qarshi State University, Qarshi, Uzbekistan

One of the urgent issues to improve the geo-ecological environment is the creation of new effective and affordable technological solutions for wastewater treatment. Full biological treatment of wastewater cannot be performed before the required standards are met, such as treatment for harmful and widespread pollution, synthetic surfactants, oil, nitrogen and phosphorus compounds, heavy metals and other solutes. Therefore, for wastewater treatment it is necessary to combine technology for biological treatment with physical and chemical methods. To this end, I have developed a chemical and biological wastewater treatment technology using raw materials which are more affordable local origin for the climate of Uzbekistan (Fig. 1).

According to the technology specifications, acid pre-treatment of wastewater in the reservoir was carried out with the help of higher aquatic plants. A collector was fabricated from Hissar mountain limestone with length L=50m and a width B=0,5m (Fig. 2).

A dimensions collector is set using a mathematical model based on the amount of incoming water. The natural composition of Hissar mountain limestone is (in %): SiO_2-5,2; TiO_2-0,05; Al_2O_3-0,8; Fe_2O_3+FeO-0,55; MnO-0,05; CaO-43,0; MgO-8 0;

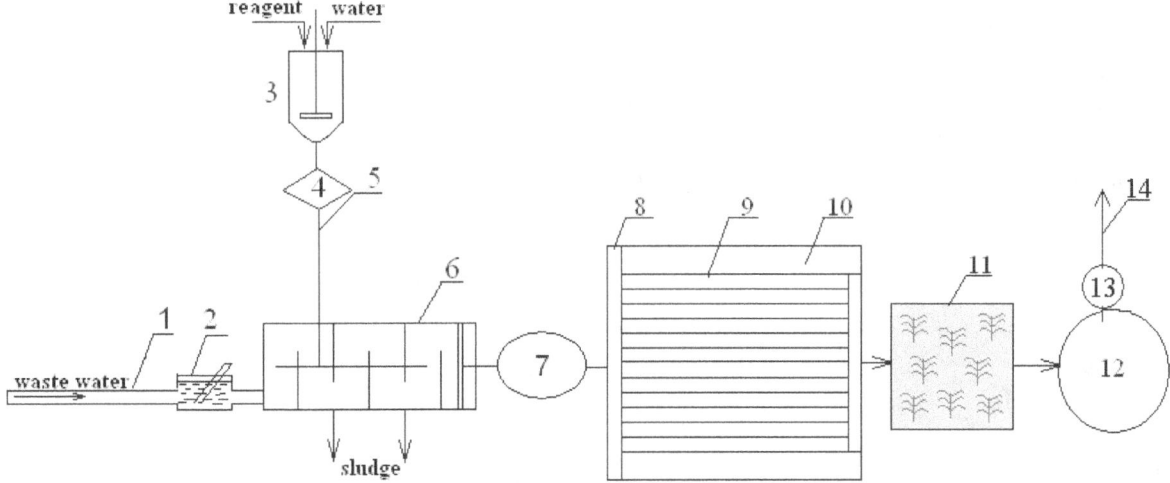

FIG. 1. Chemical- biological wastewater treatment
1-collector, 2- lattice, 3-mixer, 4-batcher, 5 - tube for feeding the solution, 6-neutralization, 7, 13 - pumping station, 8 - rasprostranitelnoy central tube , 9 - pipe network for sewage supply in bioponds 10 - bioponds , 11- gidrobioploschadka, 12 - treated water tank , 14 - conduit for reclaimed water use feedback.

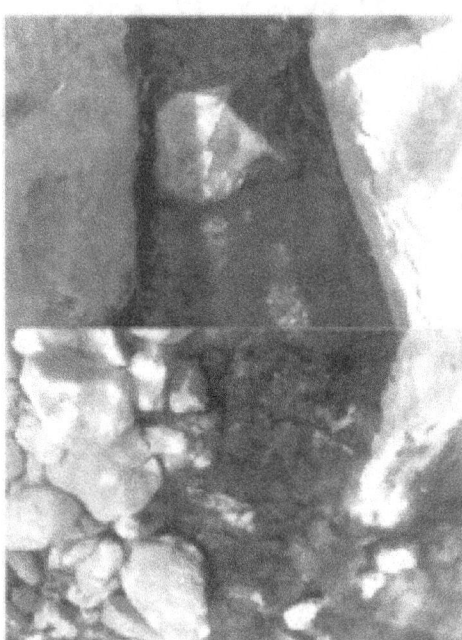

FIG. 2. Collector of limestone

FIG. 4. Type of higher aquatic plants in the device

K_2O-0,3; Na_2O-0,05; H_2O-0,75; P_2O_5-0,04; CO_2-41,5; SO_3-0,04; S-0,08 [1].

The collector wall is built of 300-350mm large pieces of limestone, and a filter ion exchange material is created at the bottom of the collector load using 30-80mm small pieces.

Layer material loading, through that filter, was only 15cm. Results showed that through the use of higher aquatic plants ans a filter loading ion exchange material of limestone in the collector, efficiency in a preliminary wastewater treatment in a short time increased by ~ 15%.

During the growing season, higher aquatic plants (under the influence of periphyton and plant rhizosphere nutrient and heavy metal ions) undergo desorption and consumed higher aquatic plants

from an aqueous solution, which provides the biological regeneration of ion exchange material.

Given the rapid within the manifold, the cultivation of higher aquatic plants used tool (Fig. 3,4).

In the collector (1) and hydrobiofied (11) for the cleaning of wastewater year-round, we used one of the local species of the higher aquatic plants family (Brassicales), race (Nasturtium), order (Brassicaseae). This species grows in wetland conditions near mountainous areas in the southern parts of Uzbekistan (Kashkadarya region). The vegetative processes in open areas lasts up to 12 months (Fig. 5).

This kind of higher aquatic plants is excellent fodder for animals, birds, and fish.

These useful plants, rich with various substances such as iron, phosphorus, potassium, iodine, nitrous oils, vitamins A, B, C, D, E, and K, contain glyukonasturtsin glycoside, saponins, alkaloids, and 3-4% carbohydrates. Its seeds contain 22-24 % fatty oil, and it is composed of oleic, linoleic, erucic, palmitic, stearic, and linoleic acid. The opportunity to get the plants on junk territories

FIG. 3. Adaptations for cultivation of higher aquatic plants in the reservoir

FIG. 5. View hydrobiofied

renders a significant harvest of 200 tons per hectare (which can be recycled for other purposes). It can be cultivated in all polluted waters and propagates vegetatively.

(rhizomes) right in the waters. A rhizome surface was arranged in horizontally dense thickets and powerful forms in terms of fault organic contaminated ponds, small lakes, and canals.

Before, deep wastewater was filtered through the lattice (2) and to extract them from the major impurities. A lattice is made of metal rods and set in the path of the wastewater at an angle of 60^0-70^0. The width of the lattice is 16–19 mm. The speed of the wastewater between the rods 0,5-1,0 m/s. Cleaning the lattices from residue of waste produced can be performed in manual mode.

After that, the wastewater enters neutralization (6) inclined type, angled at the bottom of the create favorable conditions for the deposition of impurities. This process works on the principle of Taylor vortices in the laminar regime. To neutralize the acidic water, lime water cheap solutions are used at 5-10% in which the composition comprises, in %: CaO- up to 52,0; MgO-5,0; SiO_2-28,0; CrO_3-5,5; Al_2O3- 6,0; Fe_2O_3 + FeO- 3,5. The solution is prepared in a mixer (3) in such a way that the biological cleaning of incoming wastewater with active water is done within the reaction pH=6,5-7,5 and fed through the batcher (4). A solution is prepared using a stirrer for stirring the serrated edge of liquids containing solids. This agitator separates particulate aggregates with teeth arranged on the periphery. This method can be used at high speeds without creating significant flow.

Currently, the main disadvantage of bio-ponds for wastewater is that normal operation occurs in the warmer months, and deteriorates even when water temperature is below +6°C. On further lowering the temperature, and especially after the formation of ice when the penetration of oxygen into the water does not occur, the oxidation of organic matter is almost completely stopped. To resolve this shortcoming of bioponds, I have developed a new way of operating with full analogous processes that take place in the aeration basin mixer. Treated wastewater by means of a pump (7) through the central tube rasprostranitelnaya (8, d=300mm) and the pipe network (9, d=100mm) is fed into bioponds (10). To supply treated waste water to bioponds, a plastic pipe is used, which is resistant to the action of chemical acids and alkalis.

Due to the very smooth inner surface of the plastic pipe, pressure loss in feeding the wastewater is reduced by 10-30%, compared to the metal. Over a considerable length, that can dramatically reduce the number of connections for easy installation of pipelines, as well as reduce labor and installation costs. Plastic pipes have very high moisture, which allows their use in any environment without any anti-corrosion coatings. We chose polyethylene pressure pipes that are 8-10 times lighter than steel and have high elasticity; these pipes can be joined by welding and also by using shaped parts and connecting bolts with rubber rings. For draining and mixing of sewage pipes in bioponds, the bottom opens with holes of different diameters (Fig. 6,7)

This type of aeration is more economical than pneumatic and mechanical. For dissolved oxygen

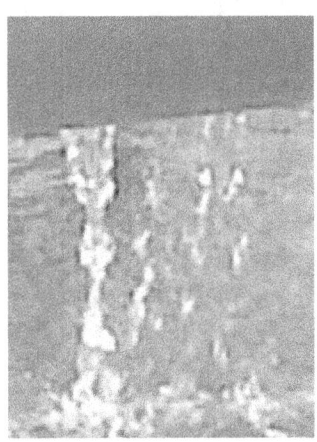

FIG. 6. The process of draining water

FIG. 7. The process of mixing water

content of treated water in the pipe, bioponds assemble complexes 1,0-1,2m above the water level in the ceramic struts, with a gap between the tubes of 1,5 -2,0m. Ponds make small depth of 1,0-1,5m, having a slope of 0,005. This allows us to easily increase the dissolved oxygen content to 70% in the winter season, which makes full use of bioponds' volume during the year. Water freezes in the winter and at the rate of oxidation little pollution reduced. Construction of bioponds is economically feasible. The advantage of this method lies in the low capital cost, the possibility of manufacturing non acid materials, and a small area for their construction. When calculating bioponds, determining their size and duration is necessary to provide them wastewater. The basis of the calculation determines the rate of oxidation, which is evaluated by the BOD and taken for a substance that decomposes more slowly.

Wastewater flowing through the pipes with a temperature +24-26°C. With all parameters, bioponds in winter in Uzbekistan do not allow for the water temperature dropping below +8°C.

There is guaranteed performance of all groups of the activated sludge microorganisms. When activating sludge which is suspended, it promotes decay of the activated sludge flows down into smaller areas and increases the rate of supply of oxygen and nutrients to the microorganisms, resulting in faster cleaning. Chemical oxygen demand of water does not increase and the so-called secondary pollution will not. Oxygen is sucked out of the atmosphere with vigorous stirring of bioponds' content. Analysis of the laboratory data shows that the dose of activated sludge in bioponds is $\alpha i=2\div3$ g/l. In the process we observed of unicellular green alga Chlorella in biological ponds.

After cleaning in bioponds (pH =6,5-7,5) for water reuse, treatment was conducted in the Gidrobotanicheskoy pad (11) with higher aquatic plants. Here was carried out at a speed of surface filtration 0,007-0,01 m/c, at a distance of 12 m, through the thickets of higher aquatic plants, with a degree of coverage of 95-100 % (Fig. 8).

The hydrobiofied playground (simple open collector) size is determined using a mathematical model, based on the requirements of the wastewater, which is cleared.

FIG. 8. Degree of coverage of higher aquatic plants in hydrobiofied

In the process of photosynthesis at the Gidrobotanicheskoy site, the higher aquatic plants saturate the water with oxygen, as well as obscure the underlying layers of water, creating an unfavorable environment for the life of blue-green algae and the formation of the primary production of phytoplankton. In the process of metabolism allocated, higher aquatic plants in active substances such as volatile production and antibiotics contribute to disinfect the water. Oxygen and metabolites in isolated plants stimulate the development and activity of periphyton living on their surface. In turn, with periphyton and mainly bacteria in the oxidation and destruction of the substances, this process is much faster. The result is a compound more readily available to plants. Higher aquatic plants are a competitor of unicellular algae and help the withdrawal from the aquatic environment of nutrients and other contaminants. This is due to the fact that these higher aquatic plants are developing very fast, therefore, consume a large amount of nutrients (pollution), exempting them from the drains while markedly changing the chemical composition and physical properties of sewage: decreased oxidation, very actively retrieval of all forms of nitrogen, phosphorus contents were missing as well as their compounds, dissolved oxygen appeared. Wastewater after cultivation of this plant on it became clear and odorless. We found that a lot of suspended solids were deposited in thickets of plants with emergent leaves, covering the surface of the water, creating a protective environment conducive to rapid settling of suspended solids in an aqueous medium. Water purification from various impurities will depend on the thickness of the width of the grass thickets through which the water flows. The length of the roots of this species of higher aquatic plants reaches 30-35cm. Using this kind of higher aquatic plants for cleaning (post-treatment) wastewater bioengineering technology is environmentally safe. Purified water is then collected in a tank (12) and is directed by means of pumps (13) and the tube (14) for reuse.

To control the quality of treated wastewater can use live indicators: Caddisflies, carp, frogs and freshwater mollusks from the family Sanguinicolidae, Diplostomidae. They instantly adjust to changes in the aquatic environment. To indicate the purity of water, caddis flies can be used since their larvae die in dirty water.

Analysis of the laboratory data for bioengineering acidic wastewater treatment shows that bacterial contamination is reduced by 90,0-99,6 % and microbial number by the number index of 99,9%. Concentration of total nitrogen is reduced to 95,0%. BOD_5=0,5-5,0 mg /l, COD=0,3-0,5mg /l, pH= 6,5-7,5, dissolved oxygen 4÷6 mg/l, suspended matter - 0,24 mg /l, total hardness < 7 mg eq/l and smell of water < 2 score.

The purified water meets all the requirements of sanitary - technical indicators for reuse in industry and technical purposes.

References

1. Khurramov M.G Secondary treatment of wastewater in the climatic conditions of Uzbekistan for reuse. - Karshi cardio 2012.- P.184.

Study of the Process of Hydrogenization of Cotton Miscella

Fayoza U. Suvanova

Qarshi Engineering and Economic Institute, Qarshi, Uzbekistan

The most of perspective directions to the technologies of working out the plant oil and fat fake the purpose of receiving solid fats for soapworks and other industries is its hydrogenization in the solvent.

Application of organic solvents permits to fall the temperature of the process of hydrogenization, to decrease the loss of the raw and prepared products. This explains that the use of the solvent permits to decrease the energy to hydrogenizing connection on the surface of the catalyst and increases the velocity of joining the hydrogen to it.

Catalysts are ones of the main components of technological systems of hydrogenization of plant oils. In addition to the number of perspective catalyst can be considered the alloy nickel-aluminum catalysts of hydrogenization of oils and fats with different additions.

The latest results dictates the necessity of improving the properties of alloy catalysts.

The analysis of famous works [1-5] showed that on the base of nickel-molibdenum-aluminum alloy, by the way of promoting, can be achieved high effective catalysts. The greatest effect is observed while using the initial alloy of little doze of the promotor. Using such metals as palladium, platinum and rhodium increases the value of nickel catalyst less than 10 persent while incteasing its activity to 50-100 percents [6].

We investigated the promoting influence of palladium and rhodium to the nickel-copper-molibdenum-aluminum catalysts and worked out new contacts for hydrogenization of oils and fats [7-9], which differ from famous high activity, stability, isomeric capacity and others (table 1).

In the table we can see that the addition to the nickel-copper-molibdenum-aluminum catalyst of the palladium and rhodium increases its activity. This increase can be observed while adding little dose of promotors. Further increase of their contain practically doesn't influence to the activity of initial catalyst or supplies its decrease.

The developments of technologies in hydrogenization of plant oils in solvents are tightly connected to specific character of oil-extraction production, where recently is appearing a wide application of solvents, received by the method of direct distillation of the oil mixture with gas condensate in the ratio of 50:50.

The certain solvent consist of a large amount of the low-boiling hydrocarbons anol comparatively little fragrant hydrocarbons. The volume of sulphureous combination is found in available borders of technical conditions of norms.

We conducted experiences on hydrogenizing the cotton oil in this solvent on the columned reactor by the flowing method. As a stationary contact, we used the alloy of the nickel-copper-molibdenum-rhodium-aluminum catalyst.

Table 2 was presented results of hydrogenization of cotton oil in low-boiling hydrocarbon solvent (50%-cotton miscella) in different conditions.

36

TABLE 1. Reference to the activity of catalysts, depending on the changes of the alloy contain

Nymler of a catalyst	Iodine amount of fats, % J_2	The activity of the catalyst (ml per an hour)
1.	66,3	0,433
2.	63,5	0,461
3.	60,7	0,489
4.	56,4	0,532
5.	55,5	0,541
6.	58,2	0,514
7.	57,8	0,518
8.	63,4	0,462
9.	60,9	0,490
10.	58,6	0,510
11	57,7	0,519
12.	62,1	0,475
13.	59,3	0,503
14.	57,2	0,524
15.	56,9	0,527
16.	58,1	0,515
17.	57,8	0,518

Seeing in Table 2, the change of the solvent nature and also transmission from traditional extractive petrol to the low-boiling hydrocarbonaceous solvent resuls in the considerable increase of hydrogenization (velocity of process). As a result, the temperature of melting and solidness of received fats increases and also adapts to with indexes a of the selectivity of hydrogenization process of 50 percent cotton miscella (S).

Hydrogenization of cotton oil in the solvent flow in low temperature (55-60⁰C). It permits to economize thermal energy and to save the quality of the received hydrogenization.

Once, when realizing these technologies, one can observe an intensive fall of the activity of hydrogenizing the catalyst, which explains its poisoning of sulphurous organic connection of the solvent.

Proceeding from the theory of Maxted, the mechanism of the action of these things, poisoning the catalyst, can explain that toxic action shows only molecules, having a pair of electrons or unused valent orbits, thanks to which can be possibe to connect them to the hydrogenizing a catalyst.

While not all sulphide containing connection is from catalysis poison.

Confirming the method of Maxted, it depends on things, connected to the surface of the catalyst with a covalent relation and the detoxication occurs in that condition when this poison (the element) will be completely changed into the generalising electronic octet.

Analogically one can estimate the influence and other forms of sulphurous organic connection on the work of hydrogenizing catalysts.

TABLE 2. Indexes of the process of hydrogenization of cotton oil in hydrocarbonaceous solvents (50%-cotton miscella)

Condition of the experience						
Tempe-rature, °C	Pressure of kPa	Volume, velocity of the presenting, hour⁻¹	Iodine amount % J_2	Temperature of melting of the fat, °C	Solidness of the fat, g/sm in 15 °C	Selectivity, (S), %
Extractive petrol						
70	300	2,0	71,0	37,3	185	81,3
80	300	2,0	69,5	38,9	195	82,5
90	300	2,0	65,2	39,5	200	85,0
90	200	2,0	70,8	36,0	200	88,0
90	400	2,0	64,9	40,5	195	73,2
90	300	1,0	51,4	50,1	370	75,0
90	300	1,5	58,6	45,0	270	85,0
90	200	2,5	76,8	34,8	190	87,7
The solvent achieved by the oil mixture with gas condense						
70	300	2,0	70,6	37,5	190	89,1
80	300	2,0	69,0	39,2	225	89,3
90	300	2,0	64,7	39,9	232	85,4
90	200	2,0	70,0	36,6	190	89,0
90	400	2,0	64,0	40,9	235	75,3
90	300	1,0	50,7	49,8	380	78,3
90	300	1,5	58,2	44,7	280	88,0
90	200	2,5	76,0	35,3	210	88,9

We conducted suitable research for the purpose of studying the influence of the certain solvent to our catalyst.

Experiences were conducted in the laboratory equipment with the stationary catalyst by the «flowing» method. In this case, the volume velocity of the presenting miczella was equal to 1,0 hours ⁻¹, the temperature of the process of the hydrogenization was kept in 80± 50⁰C, the pressure and the volume velocity of the absence of surplus hydrogen contained 300 kPa and 30 hours ⁻¹.

Hydrogenization was subjected to the refined cotton miscella with the iodine number 102,6 % J_2 and the colour 16 red number while 35 yellow in the 13,5 sm layer when containing oleic acid 20,1%; linolec acid 53,6%. The results of the experiences were presented in Table 3.

Having shown in Table 3, while using a new solvent with a little content, one can see a considerable decrease of falling the activity which is connected to its hydrogenizing with catalyst. The same components of the catalyst are more stable to these

TABLE 3. Changes of the activity of the stationary catalyst during hydrogenization of 50% cotton miscella, received in famous and offered solvents

Time of hydrogenization miscella, hour	Changing of the activity of the catalyst			
	miscella, received in the famous solvent		miscella, received in a new solvent	
	Δ l. n., % J_2	%	Δ l. n., % J_2	%
24	60,0	100,0	60,2	100,0
120	57,3	95,5	59,3	98,5
240	51,5	85,8	55,9	92,9
480	46,8	78,0	49,8	82,7
720	42,9	71,5	47,5	78,9

things, others are less stable. Thus the necessity of studying the stable activity, of wore ked stationary catalysts arises while hydrogenization of the refined cotton miscella.

We studied the stable activity of the offered catalyst №10 in the laboratory condition when the temperature is 90°C, pressure 300kPa, the volume velocity of hydrogen and 50%- refined cotton miscella 30 hours⁻¹ and 2,0 hours⁻¹.

The achieved results were presented in Table 4.

Decreasing of the activity of the catalyst were determined by the following formula:

$$\Delta A(\%) = \frac{A_{begin} - A_{end}}{A_{begin}} \cdot 100,$$

in this case ΔA- falling of the activity of the catalyst,%;

A_{begin}, A_{end} – the beginning and the ending activity of the catalyst

TABLE 4. Stable activity of the famous №1 and the offered №10 catalysts

Duration of the work, hour	Catalyst №1			Catalyst №10		
	l. n., % J_2	Activity, Δl. n. ml/hour	Falling of the activity. %	l n., % J_2	Activity, Δl. n. ml/hour	Falling of the activity, %
8	62,3	0,473	0,0	51,4	0,582	0,0
24	63,8	0,458	3,1	52,1	0,575	1,2
120	68,6	0,410	13,3	54,8	0,548	5,8
360	71,1	0,385	18,6	58,3	0,513	11,9
600	75,3	0,343	27,5	62,4	0,472	18,9
840	79,4	0,302	36,2	66,6	0,430	26,1
1080	-	-	-	71,1	0,385	33,9
1200	-	-	-	75,9	0,337	42,1

Haying described in Table 4, the offered catalyst №10 compared with the famous catalyst №1 is firmly working during 1200 hours. As a result, using the promotor of the rhodium in the nickel-copper-molibdenum-aluminum alloy increases not only its activity, but also stability.

In this form for hydrogenizing the cotton oil in the solvent for the purpose of receiving solid fats, we worked out- the melting catalysts, which are available to keep its activity during the long period.

References

1. Abdurahimov, A., Y. Kadirov, A.S. Safaev, and Y.R. Yakubov. 1976. Catalyst for the hydrogenation of oils and fats. A.S. № 539602 (the USSR). Published in the B.I. № 46.

2. Salidzhanova, V.S., F.U. Suvanova, Y. Kadirov and others. 1982. Catalyst for the hydrogenation of vegetable oils and fats. A.S. № 1010751 (the USSR). Not to be published in the open press.

3. Suvanova, F.U., S.A. Abdurahimov and Y. Kadirov. 1983. Catalyst for the hydrogenation of vegetable oils and fats. A.S. №1088185 (the USSR). Not to be published in the open press.

4. Sokolsky, D.V., K.A. Zhubanov. 1972. Hydrogenation of vegetable fats. Alma-Ata: Science, pp.181.

5. Glushenkova, A.I., A.L. Markman. 1979. Fat hydrogenation. Tashkent: Fan, pp.144.

6. Kadirov, Y., S.A. Abdurahimov, Y.R. Yakubov and others. 1976. Catalyst for the hydrogenation of vegetable oils and fats. A.S. №539602 (the USSR). Published in the B.I. № 46.

7. Mazhidov, K.Kh., A. Abdurahimov and Y. Kadirov. 1978. Cottonseed oil hydrogenation by the dead skeletal catalysts. Dep. Head in the All-Union Institute of Scientific and Technical Information № 239-78, pp.7.

8. Nazarova, I.P., F.M. Kantsepolskaya and A.I. Glushenkova. 1976. Catalyst for the fats hydrogenation. A.S. №506972 (the USSR). Published in the B.I. №10.

Research of Characteristics of Fibrous Filter Materials for the Waste Water Treatment in the Development of Placer Deposits

Kristina V. Svalova

Transbaikal State University, Chita, Russia

Abstract. *In this paper we considered the problem of wastewater treatment faced by the mining enterprises in the development of placer deposits from contaminating those suspended solids. The mechanical treatment has been improved – filtration through the introduction of new materials as the filtering partitions, namely, the needle-punched and thermally bonded polymer fabrics and filter cloths of polypropylene fibers and viscose. Herein we present you the results of studies of the given materials characteristics: a retentivity and a filtration rate. The results obtained show the expediency of using those materials in the filtering devices for the sewage treatment, particularly in the cassette type filters.*

Keywords: *waste water, filtration, retentivity, filtration rate.*

The main event to protect the surface water bodies against industrial pollution is a high quality purification of effluents. The problem of surface waters clarification from the suspended solids gets a special acuteness for the mining enterprises in the development of placer deposits of minerals, where a large volume of waters polluted with the suspended solids (clay, fine sand) is reset to the small rivers. The following mechanical methods of purification have received the greatest distribution in such conditions: sedimentation and filtration.

Filtration is the separation processes of inhomogeneous systems or suspensions by means of porous walls, which detain some phases of those systems and let pass others. Filtration is a hydrodynamic process, which rate is directly proportional to the pressure difference generated on both sides of the filter material and inversely proportional to the resistance experienced by the liquid during its motion through the pores of diaphragm and the layer of the formed sludge [1]. The waste waters of mining enterprises act herein as the suspensions and as the porous diaphragms – the filter material. The required distinctive feature of any filter diaphragm is a presence of through pores therein, which are able to infiltrate the liquid and to detain the solid particles of suspension.

The main objective of any filter device is to provide the desired clarification degree of industrial effluents [2]. This task may be performed at the successful combination of two basic parameters: a filtration rate coefficient and retentivity coefficient of the filter. Both parameters are directly dependent on the characteristics of the used filter material.

The conducted researches have shown that the application of fibrous polymeric materials is an effective, due to a number of properties that favorably distinguishes them from the traditional grainy loading, including the durability due to the use of synthetic fibers, low cost (1 m² does not exceed 50-60 rubles), high tensile strength (70-600 N/cm²), high coefficient of filtration (40-50 m/day), low weight and thickness, manufacturability and ease of use [3].

Nowadays, there is a wide range of fibrous media, but the materials listed in the Table 1 are of our greatest interest. In a detailed study of those cloths we have found that the main characteristics are the volume and surface densities for the needle-punched and thermally bonded fibrous materials, respectively, and the pore size for the fabric.

Bulk density β, $kg/m3$ can be found by Formula: (GOST 15902.2-79):

$$\beta = \frac{m}{L \times B \times h} \tag{1}$$

Surface density m_s, g/m^2 is determined by the ratio of sample weight to its area by GOST 15.902.1-80:

$$m_s = \frac{m}{L \times B} \tag{2}$$

Sp – is a pore size, is determined by experiment for each sample on the microscope of brand BMI -1. In Fig.1 a picture of the fibrous media under a microscope is shown.

In Fig. 3-5 the obtained dependences of estimated parameters are shown: the coefficients of filtration rate and retentivity for different material types.

Charts explanations (Fig. 2-5):

Coefficient of longitudinal filtration C_f, m/sec, can be found by Darcy's law [4]:

$$K_\phi = \frac{Q \times L}{F \times \Delta P} \tag{3}$$

where Q– is a liquid flow rate, m³/sec

TABLE 1. Filter materials characteristics

Name of materials											
Needle-punched nonwoven				Nonwoven with thermal bonding				Filter cloth			
Item	β, kg/m³	C₃	Cf, ·10⁻³ м/с	Item	m_s, g/m²	C₃	Cf, ·10⁻³ m/sec	Item	Sp, ·10⁻² µm	C₃	Cf, ·10⁻³ м/с
									Viscose material		
1	2	3	4	5	6	7	8	9	10	11	12
I-100	20	0,34	1,20	T-70	70	0,43	1,08	TX-30	200	0,47	1,05
I-200	40	0,41	1,18	T-100	100	0,46	1,06	TX-35	180	0,64	0,97
I-300	60	0,46	1,15	T-120	120	0,49	1,04	TX-50	160	0,65	0,90
I-400	80	0,49	1,10	T-150	150	0,53	1,01	TX-70	130	0,75	0,83
I-500	100	0,57	0,97	T-200	200	0,62	0,99	Polypropylene material			
I-600	120	0,66	0,94	T-250	250	0,66	0,95	PL-30	190	0,5	1,01
I-700	140	0,71	0,90	T-300	300	0,74	0,91	PL-35	170	0,63	0,94
I-800	150	0,77	0,86	T-350	350	0,77	0,88	PL-50	150	0,68	0,87
I-900	180	0,86	0,82	T-400	400	0,79	0,84	PL-70	120	0,76	0,80
I-1000	200	0,91	0,76	T-500	500	0,84	0,86	PL-100	100	0,81	0,74

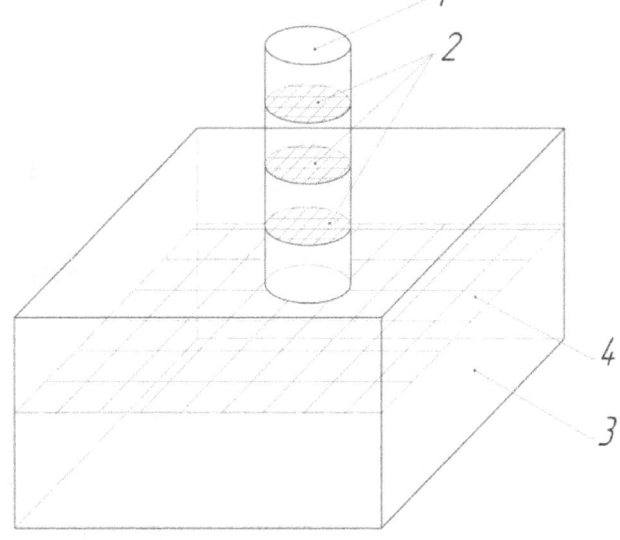

FIG. 1. View of the filter material under a microscope.

FIG. 2. Diagram of the laboratory facility for filtration.

$$Q = \frac{V}{t} \qquad (4)$$

L– length of the sample fibrous material, m

F– cross-sectional area of the sample, F= 0,00636 m^2

ΔP– pressure difference, ΔP=0,07 m

V – liquid volµme, m^3

t – filtration time, t= 60 sec

Coefficient of retentivity Cr (Кз) by the following dependence [1]:

$$K_{\mathfrak{z}} = \frac{N_1 - N_2}{N_1} \qquad (5)$$

where N_1, N_2 – the content of suspended substances in the liquid samples before and after the filter diaphragm respectively, N_1=50 g/l, N_2 was determined experimentally in the laboratory facility shown in Fig. 4.

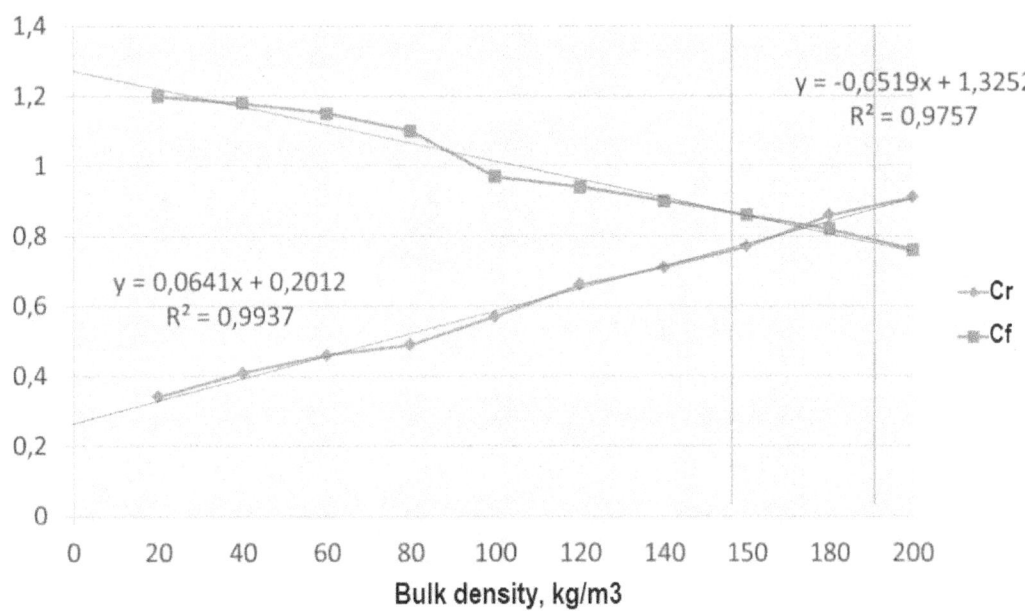

FIG. 3. Dependence of the coefficients of filtration rate (C_f) and retentivity (C_r) of the bulk density of needle-punched nonwoven materials.

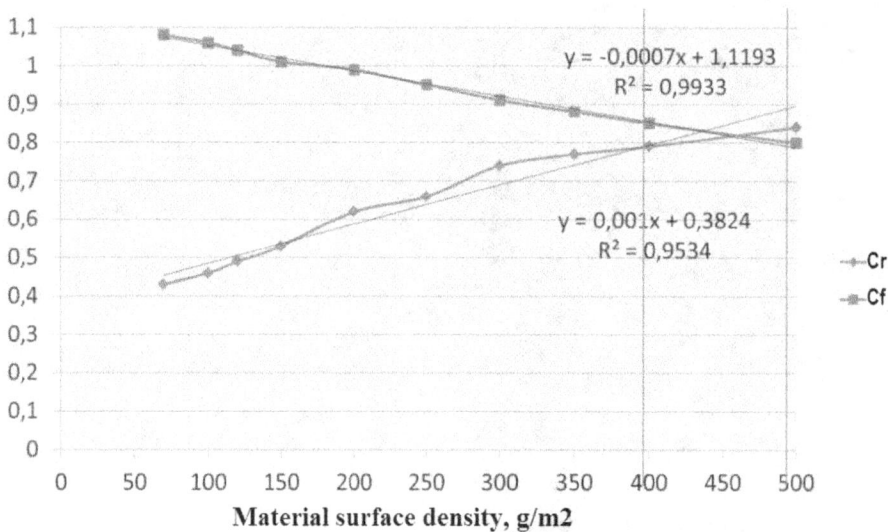

FIG. 4. Dependence of the coefficients of filtration rate (C$_f$) and retentivity (C$_r$) of the surface density of thermally bonded materials

The facility consists of a cylinder (1) of diameter 9 cm, fixed with grill (4) in the rectangular tank (3) of sizes 22*15*23 cm and volµme of 8 l. It's made of glass and transparent plastic that allows you throughout the entire experiment to monitor the filtration process and to identify its regularities. The filter material is placed on the grill (2), The filter material is placed on the grate fixed at a distance of 7 cm from the edge in the cylinder pocket. The pictorial diagram of facility is shown in Fig. 2.

We know that the material composition of the dispersed phase of waste waters include the organic and mineral substances, which are mainly consist of a mixture of nonmetallic, mainly clay minerals (50%) [6]. That is why, we have used the polluted water, obtained by mixing water with the absolutely dry clay in concentration of 50 g / 1 l of water for the laboratory tests as the liquid to be filtered.

Based on research findings (Table 1) we determined that the most optimal are the needle-punched cloths of bulk density of 150-190 kg/m³, the thermally bonded materials of surface density of 400-500 g/m², the filter cloth with a pore size of 100-120 microns (µm). The obtained results

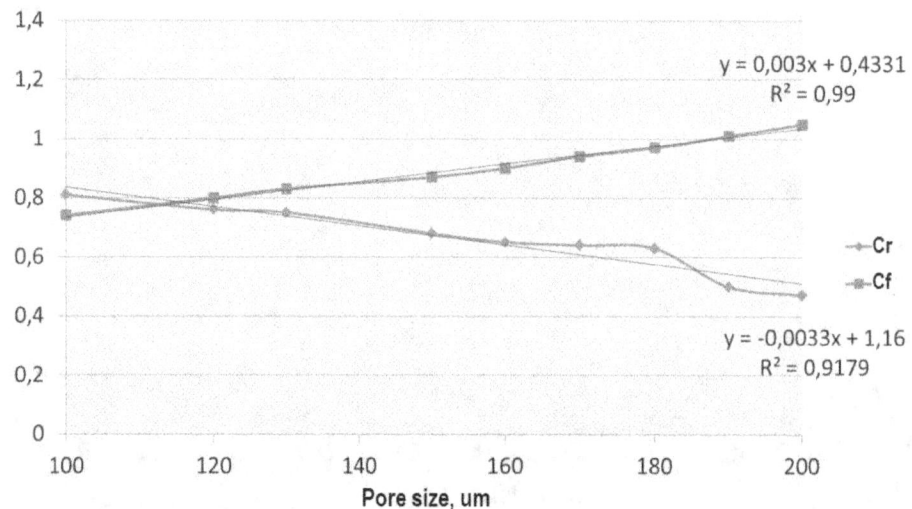

FIG. 5. Dependence of the coefficients of filtration rate (C$_f$) and retentivity (C$_r$) of the pore size of filter cloths

show the expediency of application of these materials in the filter devices for the industrial waste waters treatment, particularly in the cartridge type filters.

References

1. Zhuzhikov, V.L., 1971. Filtration. Theory and practice of suspensions separation. V.L. Zhuzhikov. – M.: Chemistry, pp: 440.

2. Abramov, N.N., 1974. Water supply. N.N. Abramov. M: Stroyizdat, pp: 480.

3. Svalova, K.V., 2013. Experimental investigations of particle retention of the solid phase during the mechanical waste water treatment by filtration using the fibrous polymeric materials. Mining informational and analytical bulletin (GIAB) # 6, pp: 391-396.

4. Gerasimov, V.M., 1998. Fibrous and film source materials technologies in the mining operations. V.M. Gerasimov, A.V. Rashkin, Chita: ChitSU, pp: 91.

Quantitation of Macro and Micro Elements of Water of North Fergana Channel

Erkin Vokhidov[1]
Saodat Gofurova[2]

[1]Namangan State University, Namangan, Uzbekistan
[2]Dustlik Academic Lyceum, Namangan, Uzbekistan

It's known there are many factors for the development and distribution of algoflora taxon in North Fergana channel. One of them is difference of micro and macro elements and its concentration in living media of algae. This factor effectively influences for the distribution of algae through the channel. The channel situated in the high populated region of Namangan region. Because of that there is high anthropogenic influence for the North Fergana channel. It was investigated mineral composition and types of macro and micro elements of the channel from the different places of the channel. The samples were obtained from the beginning and after 9 kilometers distance of the channel. The obtained results presented in the following table.

It was defined that quantitation micro and macro elements in the North of the Fergana channel were different. This data shows that water of the

TABLE 1. Quantitation of macro and micro elements at the beginning of the North Fergana channel [mkg/l, july, 2011.]

Element	Quantitation	Element	Quantitation	Element	Quantitation
Ca	58676	Mn	8,9	Cs	0,057
Na+	21414	Lu	6,06	Th	0,043
Cl-	11307	Cr	0,99	Hf	0,032
Sr	517	Rb	0,93	Au	0,019
W	1,84	As	0,86	Sc	0,018
Ba	135	La	0,37	Cu	0,065
Fe	59,5	Sb	0,32	Ag	0,01
Zn	16,5	Se	0,29	Eu	0,0085
Br	16,4	Ce	0,28	Hg	0,001
Mo	10,4	Co	0,07		

TABLE 2. Micro and macroelements quantitation in the water of the North Fergana channel after Namangan city [mkg/l, july, 2011.]

Element	Quantitation	Element	Quantitation	Element	Quantitation
Ca	81147	Mn	28,5	Cs	0,029
Na	14604	Lu	0,067	Th	0,027
Cl	11702	Cr	1,14	Hf	0,048
Sr	592	Rb	1,9	Au	0,079
W	1,99	As	0,74	Sc	0,027
Ba	148	La	0,4	Cu	0,067
Fe	81,3	Sb	0,69	Ag	0,097
Zn	29,3	Se	0,097	Eu	0,0090
Br	15,9	Ce	0,31	Hg	0,0081
Mo	13,9	Co	0,0053		

TABLE 3. Comparison of the water composition of the North Fergana channel at the beginning and after flow of Namangan city [mkg/l, july, 2011.]

Element	Difference	%	Element	Difference	%
Ca	-22471	27,69	La	+0,03	0,075
Na	+6810	31,80	Sb	+0,37	0,536
Cl	+395	3,49	Se	-0,193	0,665
Sr	+85	14,36	Ce	+0,03	0,127
W	+0,15	0,075	Co	-0,047	0,671
Ba	-13	0,78	Cs	-0,028	0,491
Fe	-21,8	0,268	Th	-0,016	0,372
Zn	+12,8	0,437	Hf	+0,016	0,333
Br	-0,5	0,030	Au	+0,060	0,759
Mo	+3,5	0,897	Sc	+0,009	0,333
Mn	+19,6	2,20	Cu	+0,002	0,03
Lu	-5,99	0,988	Ag	+0,087	0,897
Cr	+0,15	0,132	Eu	+0,0005	0,059
Rb	+0,97	0,510	Hg	+0,007	0,864
As	-0,12	0,162			

channel from Norin river is rich of soil minerals. There is 4 macro elements such ferrum, sodium, calcium, barium and 25 different micro and radio-active elements.

As above mentioned, the North Fergana channel flows from Namangan city and after 9 kilometers Turakurgan, Chust and Pop districts respectively. During the mentioned distance the composition of the water is changed by industrial wastes and population wastes. Changing of the water composition sharply increased after Namangan city. For this case, it was investigated composition of water after flow of Namangan city. Obtained data shown in the following table.

It was also investigated comparison of water composition at the beginning of the channel and the water after Namangan city. Comparison is shown in the following table.

Quantitation of the elements after the flow of Namangan city was increased. It was proved that increase in quantitation of 19 elements out of 29 [for example Na -31.08 %, St-14.36 %, Mn-2.20 %, Cl-3.49 %]. Quantitation of some elements as La, Se, Cs, Br and Co was slightly increased and percentage of Hf, Sc, Cu, Ag, Hg is not changed practically. Although there were decrease of elements quantitation after flow of Namangan city such Ca — 27.69 %, Lu-0.98 % and Th-0.016 %. As we suggest that increase and decrease of elements composition in the channel water is explained in the result of anthropogenic factors.

LIFE SCIENCE

Biodegradation of Pyridine by *Arthrobacter sp.*

Fatemat Khasaeva
Leonid Zakharchuk
Alexander Netrusov
Igor Parshikov

Moscow State University of M.V. Lomonosov,
Moscow, Russia

Abstract. *During growth of cultures of Arthrobacter sp. KM-P in a medium with a concentration of 2.5 g / L of pyridine, the pyridine was fully utilized in 24 hours. The stability of the process of biodegradation for pyridine with immobilized cells for three consecutive periodic processes was shown. The strain Arthrobacter sp. KM-P is recommended for the treatment of industrial wastewater containing pyridine.*

Keywords: biodegradation, microorganisms, pyridine, Arthrobacter

Introduction

As a result of human activities, many toxic compounds have accumulated in the environment. The main sources of pollution of the biosphere are the chemical and pharmaceutical industries. At the same time, methods of biotechnology are now widely applied [1]. The most successful strategy for biological treatment of wastewater is the use of microorganism-destructors, attached to water-insoluble carriers [2]. A serious place among pollutants is occupied by pyridine and its derivatives. The ability of microorganisms to utilize pyridine was discovered for the first time in the early twentieth century [3-5]. The bacteria that were isolated were *Aerobacter aerogenes*, *Serratia marcescens*, *Bacterium herbicola*, which grew in a medium containing 0.1-0.5% of pyridine, using it only as a source of nitrogen [6].

We isolated from the soil the bacteria *Arthrobacter* sp. KM-P, capable of use with pyridine [7], as the single source of carbon, nitrogen and energy.

The purpose of the study was to compare the ability of cells of the strain *Arthrobacter* sp. KM-P immobilized in a gel (calcium alginate) and cells in suspension to destroy pyridine .

Materials and Methods

The strain of *Arthrobacter* sp. KM-P was isolated from pyridine-contaminated soil samples [7]. For the cultivation of microorganisms a medium of the following composition (g/l) was used: KH_2PO_4 – 0.2; $MgSO_4 \cdot 7H_2O$ – 0.2; $FeSO_4 \cdot 7H_2O$ – 0.01; $CaCl_2 \cdot 2H_2O$ – 0.02; $MnSO_4 \cdot H_2O$ – 0.002; Na_2MoO_4 – 0.001; 0.5M 3-(N-morpholino)-propanesulfonic acid (MOPS) bicarbonate buffer - 1 L; pH 7.0 – 7.2. The pyridine was added in a concentration of 1.5 - 3.4 g/l. The culture was grown on a shaker at 220 rpm in 750 ml flasks containing 200 ml of medium with pyridine. Growth of the culture was assessed with the use of a nephelometer. Determination of the residue of pyridine in the medium was conducted with a Hitachi 200-20 (Japan) spectrophotometer (Japan) at λ=255 nm.

To obtain a cell suspension culture, *Arthrobacter* sp. KM-P were grown to stationary phase (18 h), which was assessed by nephelometry. The optical density of the cell suspension in this case was 0.9 units and was in accordance with a concentration of 2×10^8 cells/ml. The cells were separated from the medium by centrifugation at 6000 g for 10 min.

Cells of *Arthrobacter* sp. KM-P were immobilized in calcium alginate. To do this, the cell suspension *Arthrobacter* sp. KM-P in a volume of 100 ml was poured into 200 ml of sterile 3% sodium alginate solution in distilled water to obtain a 2% solution of sodium alginate with cells. Five hundred ml of 0.2 M solution of $CaCl_2$ in distilled water were prepared separately and sterilized at a pressure of 1 atm. The process of immobilization of bacterial cells was carried out in sterile conditions. For that, a 2% solution of sodium alginate with cells was added by drops to a flask with 0.2 M $CaCl_2$. The resulting granules of calcium alginate (size of 1-1.5 mm) with immobilized cells were kept in a 0.2 M solution of $CaCl_2$ for 10 to 12 hours. Then, the $CaCl_2$ solution was drained and the pellets were placed in 300 ml of sterile 0.9% solution of NaCl, in which immobilized cells of *Arthrobacter* sp. KM-P were kept at 4°C.

To conduct the degradation process with immobilized cells and with cells in suspension, a mineral medium with concentrations of 1.5, 2.5, 3.0 and 3.4 g/l of pyridine was used. To each of the flasks 1.0×10^9 cells of *Arthrobacter* sp. KM-P were added in the form of 8.3 ml of cell suspension or 25 ml of solution of calcium alginate granules. Samples from the flasks with an initial concentration in the medium of pyridine of 1.5 g/l were taken every 3 hours, and samples from the flasks with other concentrations of pyridine every 6 hours. Granules with immobilized cells were separated from the culture medium by filtration.

Results and Discussion

The initial concentration of pyridine was at a level of 1.5 g/l for consumption during 36 hours in a cul-

ture of *Arthrobacter* sp. KM-P immediately after its isolation from the soil [4]. During the time that the culture of *Arthrobacter* sp. KM-P (2007-2013 years) has been done, its activity has significantly increased. At the time of the study, the activity of *Arthrobacter* sp. KM-P almost doubled and during 24 hours 2.5 g/l of pyridine were assimilated. Higher concentrations of pyridine in the medium (3.0 g/l) caused the growth of the culture to slow down, and the content of pyridine in media of 3.5 g/l and over inhibited the growth of culture.

The increase the rate of cleavage of pyridine in a culture of *Arthrobacter* sp. KM-P over a lengthy period can be explained by processes of self-induction cells resulting from multiple passages on a liquid mineral medium containing higher concentrations of the pyridine.

Thus, a strain of *Arthrobacter* sp. KM-P is able to grow in a medium with a high concentration of pyridine, and completely utilize that compound. This is one of the most promising strains-destructors for pyridine. This opens up wide possibilities for the use of this organism in sewage treatment with pyridine. The use of microorganisms in sewage treatment plants under periodic and/or continuous cultivation leads to the accumulation of large amounts of biomass, requiring disposal. The use of immobilized microbial cells for wastewater treatment may eliminate the need for regular disposal of large quantities of biomass [7]. Immobilized microorganisms are protected from adverse impacts. At the same time, granules may be created in any form needed for the required case—granules, drives, fibers, pipes, etc. [8].

Calcium alginate gel was used as a carrier for immobilizing cells of *Arthrobacter* sp. KM-P. This choice is explained by the relatively mild conditions for immobilization of cells and by the possibility of providing nutrients and oxygen. The process of immobilizing cells of *Arthrobacter* sp. KM-P was monitored under the microscope.

The speed of utilization of pyridine by immobilized cells of *Arthrobacter* sp. KM-P was compared with the speed of utilization of pyridine in a suspension of cells. It was established that cells in suspension have a faster speed of utilization of pyridine than immobilized cells. Pyridine at the concentration of 1.5 g/l was fully utilized by cells in suspension in 6 hours. Immobilized cells assimilate such a concentration in 9 hours. Concentration of pyridine of 2.5 g/l is consumed in 12 hours by suspended cells, and in 18 hours by cells immobilized in calcium alginate. Pyridine in concentration of 3.0 g/l was utilized in 18 hours by suspended cells; pyridine in concentration of 3.4 g/l was utilized in 24 hours. Immobilized cells of *Arthrobacter* sp. KM-P use the same concentrations of pyridine (3.0 and 3.4 g/l) more slowly than cells in suspension (30 and 32 hours, respectively).

The results of the study suggest the possibility in principle of using immobilized cells of *Arthrobacter* sp. KM-P in calcium alginate for the removal of pyridine from wastewater. However, the catalyst system with immobilized cells should provide technological and economic advantages compared with microbiological processes on the basis of suspension cells. Application of the benefits of immobilized cells is only possible when used in bioreactors with batch or continuous mode [7]. In this case, in the reactor it is possible to use higher concentrations of immobilized cells of *Arthrobacter* sp. KM-P. For operation with high concentrations of aerobic bacteria in a fermenter, it is necessary to create high-speed mass transferal with the help of remixing of the catalyst pellets. The mechanical strength of granules of calcium alginate allows for such goals [9, 14].

The most promising approach may to be the model of a flow-through reactor mixing granules of calcium alginate, a variant of the continuous cultivation of independent microbial cells—chemostat. Creation of a bioreactor based on immobilized cells of *Arthrobacter* sp. KM-P for the treatment of wastewater containing pyridine is the next stage of our work.

References

1. Parshikov I.A., Sutherland J.B. Microbial transformations of antimicrobial quinolones and related drugs. // J. Ind. Microbiol. Biotechnol. 2012. V.39. N 12. P. 1731-1740.

2. Beshay U., Abd-El-Haleem D., Moawad H., Zaki S. Phenol biodegradation by free and immobilized *Acinetobacter*. // Biotechnol. Lett. 2002. V.24. P.1295-1297.

3. Parshikov I.A., Netrusov A.I., Sutherland J.B. Microbial transformation of azaarenes and potential uses in pharmaceutical synthesis. Appl. Microbiol. Biotecnol. 2012, v.95, N 4, P.871-879.

4. Parshikov I.A., Terentyev P.B., Modyanova L.V., Duduchava M.R., Dovgilevich E.V., Butakoff K.A. Microbiological Transformation of 9-Amino-1,2,3,4,5,6,7,8-octahydroacridine. Cheminform. 2010, v.26, N 10.

5. Modyanova L.V., Duduchava M.R., Piskunkova N.F., Grishina G.V., Terent'ev P.B., Parshikov I.A. Microbiological Transformation of Piperidine and Pyridine Derivatives. // Cheminform. 2010. V.31, N 12.

6. Moore F.W. The utilization of pyridine by microorganisms. // Journal of general microbiology. 1949. V.3. P. 143-147.

7. Khasaeva F., Vasilyuk N., Terentyev P., Troshina M., Lebedev A. A novel soil bacterial strain degrading pyridines. // Environ. Chem. Lett. 2011. V.9. N 3. P.439-445.

8. Park, J.K., Chang, H.N. Microencapsulation of microbial cells. // Biotechnol. Advances. 2000. V.18. P. 303-319.

9. Liu R, Shen F. Impacts of main factors on bioethanol fermentation from stalk juice of sweet sorghum by immobilized *Saccharomyces cerevisiae*. // Bioresource Technol. 2008. V.99. P. 847-854.

10. Plieva F.M., Galaev I.Y., Noppe W., Mattiasson B. Cryogel application in microbiology. //Trends in Microbiology. 2008. V. 16. № 11. P. 543-551.

11. Leenheer J.A., Stuber H.A. Migration through soil of organic solutes in an oil-shale process water. // Env. Sci Technol. 1981. V. 15. P. 1467-1475.

12. Sims G.K., Sommers L.E. Degradation of pyridine by a newly isolated denitrifying bacterium // J. Environ. Qual. 1985. V. 14. P. 580-584.

13. Sims G.K., Sommers L.E., Conopka A. Degradation of pyridine by *Micrococcus luteus*, isolated from soil. // Appl. Environ. Microbiol.. 1986. V. 51. N 5. P. 963-968.

14. Watson G.K., Gain R.B. Metabolic pathway of pyridine biodegradation by soil bacteria // J. Biochem. 1975. V. 146. N 1. P. 157-172.